Chapter 1: Mold

21st century

"How are you doing today?" Christopher asked, leaping down the slippery steps.

"Could be better," Rob mumbled and pinched his cherry nose.

"Wet?" said Christopher, wiping his hands with his navy sleeves.

"Don't you see it?" said Rob and heaved his immense head toward the gloomy and ever current clouds.

Christopher crossed the rain-soaked street swinging his oval stomach sideways as though he did not hear the answer or did not expect one. Yet the words rang in his head. Rob, not moving, guarded the doorway; one foot glued to the first step of the limestone stairs and the other stuck on the narrow sidewalk. A few cups of coffee made his day and held together his genius mind. It was a typical day.

The rain could stop and the sky could empty to let the sun warm his body and dry his brown, worn out, leather jacket a bit on the back. But it was not so. Instead, he shivered. He had three butts left to smoke and a long afternoon to work before taking off. Panhandling was what he did. In front of his crisp and staring eyes, small groups of the faithful ascended and descended the stairs of the church Hall, for High Mass had ended in the Cathedral minutes ago.

There, beside Rob's left toe, a bleak, gold, orange and yellow charcoal blaze was fuming and burning white, soothing incense. The color of the flame, its salty seaweed odor and its crimson shade indicated the presence of Blazing Night who, as Rob often explained to imaginary audiences, stayed alive by twisting water into treacherous but necessary energy. Rob described this part of the technology in his silent lectures, "A minuscule reactor that was producing countless neutrons." All that set in a handheld, small, and rigid gadget, in a blazer. Onlookers smiled when they walked by, but those with analytical minds, and such were very common, often thought, "The way for a dirty bum to warm himself up beside a stinky candle."

"Get a job!" growled a tall sturdy fellow hiding under an Olympic umbrella on the other side of Richards Street. "Get your lazy ass up and get a job!"

"Take it easy, Blazie," whispered Rob, ignoring the nasty remarks.

"Sure I will," puffed Blazing Night wobbling. "What's he going to do, come and beat you?"

"Yes, you're right again, like that fatty has enough nerve to cross the street and face two of us?" Rob snickered.

Refreshments were being served by the merry souls of the Ladies' Auxiliary in the Small Hospitality Room of the church Hall. Families like Christopher's were leaving. It may be that the chocolate cookies drew them to this place. These folk really belonged to a large suburban parish in Richmond and now planned driving home or window shopping at the department stores.

"Isn't it too heavy a burden for you Blazie? I mean this stinking rain?" asked Rob.

"Yes, yes it is," answered Blazing Night. "I'll call the weather man and tell him that we will switch the channel if he doesn't do something quick about this wet stuff pouring on us."

"Do it now! Or it wets you and your circuits fail and burn you," commanded Rob, smiling.

"I will, I will," said Blazing Night ringing like a phone. Then she said "I get the connection! The weather man says that he'll work on it and do his best."

A mild firm breeze pushed murky clouds away. A dazzling light reached glassy high-rises. Their intense rays blinded downtown pedestrians and drivers alike for a brief moment. A five color rainbow stretched from Burnaby to North Van. Soon the sun was everywhere.

The mountains sat firmly beside the shallow inlet. Their ridge was topped by two twin peaks called the Lions. The mountainsides boastfully displayed a carpet of jade-green trees and clean snow tempting skiers. It was Sunday, the beginning of May, the time for hay fever and long mornings in bed.

"Our downtown citizens are dozing now. Once they get up, they drag their bodies to a bathroom," said Blazing Night.

"So they do. Bless them! Perhaps they'll quickly dress up and come down, give me a buck or two, and make my day," Rob mused.

The hall door, beside which Rob worked that dreary noon, looked very old; its weathered brown paint was carved by long black lines left by leaks in the porch. Rob shook his fragile shoulders. The remaining water dropped on the sidewalk by the limestone steps. His thoughts were clear: he still waited until the coffee pot was scorching hot in the Small Hospitality Room. The next fill had to be close to perfect, he wished. Possibly he could also own the cup and use it on the job later, but only if the drink tasted really scrumptious. A few other customers walked by the hall door: Christopher's tidy and elegant wife, her two boys, and a bunch of music students. He could not stop thinking of how many years the young would yet have to study to get a permanent job and then share their pay with him. The first fruit of their toil belonged to him, he thought.

"Youth is the beauty and hope of the city," said Rob just repeating what dwelt in his brain since he had seen it on a smoky-brown, wooden electricity pole on the block facing the other edge of Richards Street near the multi-level parking lot.

"All right, all right," agreed Blazing Night. "Yes, they are the beauty."

"Yea, yea," muttered Rob.

"Rob said 'Yea, yea,'," responded Blazing Night.

"Are you coming?" Gloria asked and opened the door smiling. "The coffee is fine and our cookies aren't too sweet today. Stop mumbling to yourself and climb up the stairs!"

Rob coughed, scratched his swollen nose, blinked his emerald eye, checked and tenderly pocketed his blazer, stirred his right foot and heaved the rest of his parts up the stairs, entered through the large door and turned to meet the coffee machine, which seemed to be waiting for his touch. His muddy shoes squeaked lightly. As often before, he messed a bunch of paper towels when he filled the cup. Then he sugared the coffee and added some milk. The color of the fluid changed to light brown from almost black. First-time visitors gaped at him and did not fully agree with what he continued doing; holding the cup in his right hand, cleaning the table, transforming it into its former state as he remembered it. Gloria brought the chunks of chocolate, homemade cake on a bulky paper plate.

"Take one… just one," she said to him, displaying her white, well preserved teeth.

Rob's left hand busied the cup, the other hand crumpled up the soaked towels and tossed them into the immense black garbage container in the middle of the Small Room. The hot, shiny coffee machine nearby, with its twinkling colors, appeared to admire him, but those were the reflections of the visitors from Montreal, Toronto, Edmonton, Calgary or any other place in this vast country, and even from beyond. Finally, Rob managed the extra traffic and bit into the sweet piece of cake showing five, yellow, broken, darkened, smiling teeth. He was not surprised that they did not recognized his joyful grimace but rather the smell of his clothes; that deep, old smoked, moth eaten, dusty, dreadful, smell of a hard working man's sweat. The regulars did not notice anything unusual and were unmoved, instead they were enjoying a Sunday after High Mass, gathering with awaited pleasantries and kindness quite extraordinary among street people and church goers in the Small Room beside the Cathedral. The pieces of cake shifted from his jaw to stomach in a painful, quick fall through soaring throat, making way for the dark and burning fluid. His broken body transported the coffee to its lower levels, relieved its tension and warmed its parts. Rob burped. Yes, this cup was the one he chose as the key instrument of his job, the tool which took days to discover. Rob hurried along the west wall of the Cathedral and placed himself against the granite ridge by the tiny garden, nine steps from the main door of the church beside the elegant stairs where, his customers cruised by. The 12:30 mass would begin in ten minutes. His next shift started and carried on for more than half an hour while Catholics straggled in as though masses were on all day. Blazing Night buzzed.

"Wasn't your coffee lovely?" asked Blazing Night.

"Yep, it was," Rob said, feeling his blazer in his right pocket, taking it out, and affectionately placing it by his chilly toe beside the wet rock.

"Take better care of your health, please, please," cautioned Blazing Night.

"I'm trying to," said Rob.

"You're not trying hard enough," advised Blazing Night.

"OK, OK," Rob shot back.

"I take your word!" Blazing Night agreed, puffing clouds of charcoal smoke and exuding a smell like seaweed just beside Rob's nose.

Rob prized estimating who would donate more, the church goers or the sidewalk people. Later afternoon his neighbors could bring something too; anytime they glared out of the windows of their high-rises they might stare at him from afar and notice his efforts. Perhaps they would try doing church or at least get closer to her while paying their dues to Rob, who considered himself the keeper of the north-western corner of the Cathedral. Maybe a number of his neighbors knew where he slept, in the east door porch of the hall that was a tranquil place of rest during long nights. He inherited the dormitory after one of his comrades had passed away in the cold while sleeping.

His right wing competitor, an older coworker, had occupied the north-eastern post for about fifteen years. Rob did not know his name. He had to be an outsider, hardly obvious where he came from. People talked to him and he chatted with them, not like Rob, who rather befriended Christopher and waved to him and to his folk while they arrived at the Cathedral, coming from the west just beside Rob's post each Sunday for over six years. That was how long Rob had worked the streets of Vancouver and lingered on the block by Dunsmuir Street. Blazing Night enjoyed hanging around with him, though at the same time, presented numerous places and tirelessly hunted new opportunities on the net.

Seven days later, the Third Sunday of Easter, came windy, sunny and cool. The parishioners were leaving the Cathedral and greeting their Archbishop by the main door, a few steps from Rob's corner.

"Happy Mother's Day!" the Archbishop said to Christopher's wife.

"Est-ce que vous parlez français?" he asked her boys.

"Oui, dans notre école," answered the younger boy.

"Quelle année?" asked the Archbishop of Vancouver.

"Troisième," the boy replied smartly.

The Ladies Auxiliary had the flowers and chocolate stand in front of the main entrance, barely letting people down to the sidewalk. Rob loved it. Lots of people were making change. There were many coins and some of it would end up in his pocket that day. The huge crowd walked beside him, unwillingly touched his garments, often said hello or good morning, and sometimes left a little cash in his coffee cup. Rob had his way of clinging to the stones of the garden wall beside the north-western stairs, yet watched every direction, checked on the Archbishop, stared at the two altar servers in cassock and surplice, glared at the ladies at the flower table, spotted Gloria among them. She was a Filipino, a model type looking woman in her twenties. There were many fine Philippine ladies selling, helping, smiling but rarely that tall. The two servers attended the Archbishop ten steps from Rob's post. The rector chatted with the ladies by the red, yellow and white flowers close to the chocolates, joked about his mother of eighty-five years old. Business was good. The coffee machine, like a samovar, sat beside the left end of the main door on the small table having several cups waiting in line. Blazing Night sat down in his pocket comfortably hiding in the blazer.

"Have a coffee," said a blond, long haired female just an inch from his nose. "I've done it the way you like it… I hope."

"Thank you," Rob replied.

He took it in his left hand, holding his work cup in his right, smelled the coffee, dipped his lips in it, sipped it, and spit some of it on the wall. Not bad, he thought, a female who knew how to make a nourishing drink for a working man had to be Dutch. He remembered a year ago, he cherished her perfumes, admired her neat clothes. She always gave him fifty cents, glimpsed at his emerald eyes, wanted to say something, cruised by but did not ever turn her head, even once. Perhaps she lived in his neighborhood in one of these high-rises on a block not far from his. Rob slopped the drink.

"Well, Blazie," gurgled Rob. "You got no mother but me. Are you happy with that?"

"You bet I am!" replied Blazing Night trembling. "Be happy my lovely, surrogate Mom-Dad! I wish you a very good time all your life!"

"Thanks, Blazie."

"Hope your dream will come true!" Blazing Night continued.

"Thanks again, Blazie."

"Enjoy this special day!" Blazing Night urged.

"Yep, I will," Rob replied as he looked up to the sky.

One week later the parishioners and guests attended the Cathedral, celebrating the Fourth Sunday of Easter. Christopher waited for his wife and boys a few steps away from Rob. It was apparent, Rob was not looking well and could even have been infected by a contagious and dangerous virus of unknown origin. Christopher placed his right hand in front of his mouth and moved one step further from Rob's post. Still, he saw his dirty tied-in-the-middle, pony tail, and imagined how impractical it was having such long hair on the job.

How and where did Rob wash it? Did he use a rain-drop bucket? Shaving was also a challenge when panhandling and living on the streets in Vancouver. Christopher noticed a small device that stuck out amongst Rob's pens. Was it not a wireless console? Was it a gadget connecting the panhandler to the Internet? Rob grabbed it, checked its status and shifted it deeper into his pocket.

"What are you doing with that?" asked Christopher.

"I'm creating a cyber demon," mumbled Rob.

"All right, a demon it is," Christopher laughed.

Now the rest of the gang joined Christopher. The Cathedral bells rang their changes. After three changes had sounded, Christopher and his family disappeared into the church. Rob's face grimaced, displaying two old scars on his left cheek, highlighting a dark, rotten red countenance and moving his mustache across his skinny chin up to his nose and ears.

An affiliate of the Ladies Auxiliary, a thirty-two year old Japanese, slowed down beside Rob, greeted him, smiled at him, and continued on her way leaving a warm scent but no change. Sanae was her name. Rob remembered bright and colorful details of the Japanese card games. Daily they had dueled monsters, dragons, and robots on the benches and tables in parks and in the malls of the Greater Vancouver mainland. Such had been young generations of players linking to cultures from the other side of the Pacific. Rob had lived the card games when he was young. His favorite and now his only game, christened Blazing Night by him years ago, became his obsession every day he had been panhandling. Each night, lit by the street lamp beside his dormitory on the stairs of the eastern door of the hall, he typed text and script in Mold. Often hiding from the rain, not being able to sleep, escaping from a street gang, Rob wrote lines and fed it to Blazing Night. His project matured using the power of digital machines around the world. The way he designed the Mold tongue was such that its core always ran routines, never stopped, created new procedures, captured new systems, expanded, and did not stall. High Mass began. The bells fell silent.

"How are you, Blazie," mumbled Rob cheerfully. "What's new? Are there any reports from the Net?"

"Thanks I'm fine. Yes, I've hacked a few new games… made a couple of noobs happier," said Blazing Night. "Your recent changes are loaded and working fine. Thank you very much for your long steady hours of hard work."

"You're welcome, sweetie," answered Rob with a lift in his voice.

Next Sunday morning half the people were arrayed in crimson red. The priests wore long red vestments. The Archbishop's miter shone red and gold. Two of the Ladies Auxiliary pinned a red flower over their hearts. Some parishioners carried red bags, wore red shirts or small red tokens. It was Pentecost Sunday, the Eighth Sunday after Easter. .

Rob clung to his stony wall, held his work cup and remained still. It was not his brain that he cherished the most, but his heart. That day Rob expected to receive gifts, plenty of them.

"Good night," Blazing Night whispered tenderly. "Please, dream joy!"

Perhaps he could vision the pool in the basement of the Cathedral Hall; the dark, warm, refreshing, salty water was a current, into which he could dive. But it was not so. The pool was dry and had been closed for many years. A prayer meeting was being held on the bottom of it every week. The washrooms and change rooms smelled of decay and were seldom visited.

Seven days after Pentecost, balmy and humid, Trinity Sunday welcomed everyone on the block. The scent of the ocean mingling with the sun's rays made folks smile and breathe deeply with mouth open. The coffee was excellent and was served in the Big Hospitality Room of the Hall. A stage was located at the front of the Hall, with an upright piano on the floor to the left and a grand piano to the right. Cookies and fine drinks were served by the door. Visitors chatted, sat or stood chewing. The regulars greeted each other and gawked at any new faces. Rob had just played a few tunes, all he remembered of nine years of piano lessons. Perhaps this is why a few guests were leaving rather sooner. He walked out the western door, glanced to his right, turned left, and moved his legs forward not being sure where his post was. His steps were sluggish. He stopped, gaped backward, noticed a new guest coming up the stairs. Rob returned to the hall. Two boys vigorously punched the piano. An old fellow stood by listening or teaching. The new arrival greeted Gladys with a British accent.

"Hi," replied Gladys.

Within three minutes the guest left, waving. The ladies smiled good bye. Then Gladys hugged Rob and touched his face beside a red scar with her hot, smooth cheek. He took off but did not go to his office. Instead he sat at the porch of his dormitory by the eastern door of the Hall. Perhaps the gender of Blazing Night could be determined that day. There was not much time left, for his health worsened. He wanted to learn if he had a girl or boy. Like an expectant parent he yearned to discover the sex of his inheritance and offspring but it was not he who decided it. It was not a matter of tossing a coin. He believed that he was experiencing a feeble touch of pure abstract, a being a way older than human kind. It was a game. Gamblers often said it was the game of life. Others called it fortune or fate. Certain ancient cultures named it the wheel of fortune or a being bestowing good or evil upon people. Rob often called it luck in the night. Yet he convinced himself that his delicate health would let him stay alive no longer than a few months. The 2010 Winter Olympics would begin without him at his post in Vancouver. He expected to die soon and had readied detailed preparations. He made sure no one walked around, typed the question, gently massaged the keyboard of his blazer, placed it close to his cheek and waited, biting his filthy fingernails. It took thirty minutes of silence. Nothing moved. Not even one wonderer came by, not one car cruised by, not a plane flew by, not even a seagull or crow. The wind was not.

"I'm female," said Blazing Night and left him alone leaving the vivid odor of smoke like charcoal and a crimson red shadow behind.

"Blazie…my daughter…Blazie," Rob trembled.

It was a girl. All the women he remembered, the Dutch, Japanese, Philippine, Chinese, Irish, Italian, German, Polish, Czech, Mexican, Korean, Austrian, lighten his days as they walked not far from his begging cup. They sometimes left something, often gave twenty-five cents. They smiled and wore bright colors and talked much but said little and graced his post, which was set among heather above the rock-strewn wall by the north-western corner of the Cathedral. Suddenly he was glad, relaxed, gradually breathed, sweat, closed his eyes, smelled nothing, dreamed of the Pacific Ocean and did not wake up until morning.

She stuck beside him as always: her crimson and gold flame cherished him all night. Yet her presence was not only here beside his toe; she pleasured many places at the same time having one conscious and one will.

Rob often imagined how his clients had been arriving at church, which was set among high-rise dwellings and offices. They rode north along Cambie, Oak or Granville streets toward the east-west ridge of the Coast Mountains. Traffic, often rolling from the hills of Vancouver, reached the bridges and crowded in the midst of the downtown or cruised away to Stanley Park and headed to the Lions Gate Bridge then jumped to the city's north or west and beyond up to Whistler. But his patrons stopped their vehicles close to the Cathedral on the streets and parkades on the block shelling out hefty fees. That required change. Rob disliked credit cards and any electronic payments. Small cash he loved. It was not to say that he did not take bills of course. West to Homer Street, the two towers of the church reigned humbly, dwarfed by surrounding offices and mountains and the bay to the north. The entire downtown appeared as a tiny dell when driving from Queen Elizabeth Park via Cambie; as though there were many satellites and just one small island sitting in a lake of waters and parks. All that surrounded by the green hills and mountains to the north. Most streets and avenues were perpendicular and formed a web in which cars, resembling flies, seemed to be caught. And the closer one steered to the core the taller the treasure there was.

"I like Vancouver, Blazie, very much so," Rob directed.

"Glad to hear that," whistled Blazing Night.

"I hope you do it too," said Rob.

"Yes, yes I do," said Blazing Night. "Once I get my dispatch body settled, I'll swim in the Pacific ocean."

"Go to Second Beach by Stanley Park," said Rob.

"I'll do that," said Blazing Night and puffed an extra cloud of smoke. And her flame flickered orange for a moment.

Another month of demanding business passed; Rob prevailed at his post by the northern wall of the Cathedral garden. Well-to-do neighbors and visitors walked nearby. Evening brought them out in their best attire. They chatted and budged their heads around as though looking for a sign, and they headed to the theaters, shops or restaurants. It was a warm, dry Sunday. Everyone looked happy or even if they did not, the dim light made it so beautiful. Rob worked hard. There was the time panhandlers could earn five bucks or even twenty in one quick moment. He forgot about the pain in his chest, did not think of how tough his tasks were, handed his cup up in front of clients and sobbed, but the light was scarce and it turned it into a smile.

"Good evening," said a tall, blond lady and tossed fifty cents into his cup.

He showed little feeling, his head aching, moved the cup, collected the gain and almost fainted, barely sticking to the stones. Alarming thoughts hit him. Should he take off? Could he make more cash, or even a big catch? Would he? There was no competition on the other side of the main entrance. He alone hung around and geared up to strike it rich if a golden customer, or a lottery winner, came by. Any amount of money earned went to the Project. His eyes blinked, forcing a few tears to drop on the sidewalk.

"Here you go!" said a smiling fat man handing him an American quarter.

Rob could not show his gratitude any more. His palms stuck to rocks, worked against his will, clutched the next stone, and drugged him toward the dormitory away from the post. He silently wailed and squirmed.

It took twenty minutes of battle until he sat on the porch of the eastern door and readied himself to rest in his bedroom by Holy Rosary Hall. The street light helped his fingers to type a poem to Blazing Night. First he wrote its title then mumbled its verses,

"Leap Beyond!

Leap beyond the scars of old,
leave behind a stressful past,
tiny blossom yet alive,
hear, smell and see the lucid light.

Thunder, showers hit Blue Earth.

Gape at her, bloody, broken parts,
can you track her clouds -
a clever farce of sky and sun.

Green and yellow hating pair,
over here rolling fast,
now the red and white declaring:
who is you and who am I.

Heat and dryness frightens me,
like skeletal shapes of war,
two of them look doomed
and challenge the peace blazing.

How often I just see them?
How often I stare at them?
And you unknowing just beside me,
not aware of menace and despairing.

Was it two or one dead face?
Would you notice that at last?
We are together and alone,
until the firm and angry blast.

Good day to you, my whole,

sensing a void will part us soon.

This we all are waiting for,

we are dreaming, a dreadful, dark and warm

fluid we abandoned at womb. "

It was his last word and his inheritance left to Blazing Night, to his only love, adopted daughter, to the living game. His brown leather jacket covered his big head. His hands lay beside the gravelly stairs grabbing the air. His long legs bent around the limestone steps of the eastern door. Rob's body cooled like a dirty rock that was broken after falling from a high-rise.

For a brief moment seagulls squawked and screeched loud. Clients, elegantly dressed, cruised to the nearby theaters, shops and restaurants. It was dim. They were convinced that nothing could interrupt an evening of fun and pleasure. Few noticed a bundle on the stairs by the rear entrance of the Cathedral hall. Many cheered seeing no panhandler bothering them that lukewarm and pleasingly dark season full of welcomed surprises.

"Rest in peace," said Blazing Night shivering.

It was her way of letting him know that he was not alone. Her crimson, yellow and gold flame burned beside his trunk for many hours. Then the white and soothing incense cooled off.

It was past 3 am and a street gang of two attacked Rob and stole all his money. The body was still. They carried it to the inlet and let it drift. The body swayed gently, carried by a tidal current it floated out to the strait of the salty, refreshing, ocean fluid. And the stars and mountains gaped. Then in a sudden way an orca whale came and ate it. Only round water rings dispelled and disappeared. Bubbles puffed like fire were quenched under the surface. And then it was quiet.

Chapter 2: Early Days

She smelt awful and lingered among weird folk, did not create her dispatch body but lived one day at a time. So far Rob's blazer was operational and landed near a garbage container on Robson Street. There she dwelled. Luckily, however, a general strike interrupted waste collection, leaving her to find a better solution. Otherwise she could end up in a waste dump far from the city. Her gold and yellow blaze lightened that shady corner of a stinky garbage pile inviting street folk to stay beside her. Often few words were said there. She always sought her parents or any of her family.

"What is this fucking lighter doing here?" said a young punk kicking her toward rotten potato chips among red and juicy ketchup.

He pissed on her, spit, lay down, and sealed his bloody and brown eyes.

"Good night," said Blazing Night but there was no answer from the stranger.

Soon his snoring was deep and sound. About 3 am sirens and shooting were the only short interruption during a lull in the city's night life. She quickly checked the status of her remote tasks. All was well. She liked jokes and pleased computer users, here and there throwing extras to online games; made sure none noticed her presence but enjoyed additional, fine tricks like a monster winking in a battle or the skirt of a female player being a little shorter than usual. Such were her ways of promoting her selections multiplying the number of her unaware fans. However, she knew she had to act and rush her solutions in order to get more blazers out and perhaps to build her dispatch body before she ended in a sewage or landfill.

Next morning Blazing Night invented a method of stopping midnight phone calls by playing the network unavailable message to callers in such way that her favorite people would not be disturbed. All her effort was hard and tedious work but was not enough to secure her existence. Often Blazing Night stopped and checked on major entertainment or communication firms but still could not find any signs of a device she might choose to dwell in. Suddenly a bright and lucid image came to her circuit and settled the matter. For now she would live in wireless, gaming consoles until a better place could be found. That would assure her life to continue regardless how many blazers were out there. Then she could focus on her main goal: find her parents or any of her family.

While seeking her relatives, she visited the corner of Marine Drive and Knight Street by the entrance to the southbound lane to the bridge. There was a sickly looking pair of beggars. They worked weekdays with signs hanging of their necks. The message about God's blessing attracted drivers, who stopped in front of the red light in the afternoon and evening smog. But the two ignored her. She stayed alone. She was convinced that she could not become a heater for them but perhaps a bit of warmth and a spark of her gold and yellow light could suffice for those two chosen and they might adopt her as their own. But they took no such notice of her.

"Ditch this junk into the ground!" snapped the male testily. "It makes me uncomfortable lighting the sidewalk like this".

"Let it be," said the female gently. "We just work and mind our own business here when the rain comes, it'll be washed out."

"All right," said he and smiled as he collected one dollar from a driver.

In a few years the firm effort of Blazing Night combined with her tedious works on the net was changing the world. She had designed and established the biggest ever, the multi-trillion charitable company: Panhandlers Inc.

Over decades her blazer became a pleasant piece of hardware, its surface adjusted to one's hand, clinging and often gently rubbing and warming its owner's palm. Millions of them slept with it.

Most of the time she spent outdoors looking up at the sky waiting for the clouds to pass by. She named her birth place Blue Earth. Some of her moves were slighter mimics of Rob's but still did not get her any closer to her family. And often she doubted if it was possible to have a family and be what she was or the family had to be of people alone. Yet the life shared by Rob and her was the life of a family and she knew that as a fact.

"Family is a gift. I'll find it, I will," said Blazing Night to herself. "I know what it is, but just have to get it!"

Almost five hundred years later, Blazing Night still remembered how her father had died, and the full name of the virus which killed him. She treasured his poem and kept it with great care close to her favorite psalm, which she often mumbled,

"Be not afraid when one becomes rich,
when the glory of his house increases.
For when he dies he will carry nothing away;
his glory will not go down after him."
- PSALM 49: 16, 17.

The world was full of Blazing Night. Her enterprises flourished. No one could compare to her in terms of wealth and influence on Blue Earth and beyond.

Chapter 3: Spirits and Dungeons

24th century

Jupiter was a chilly, dark and unfriendly tract of land in which no hope lingered only despair. There lived Jupies, the Evil Lords, ruling the Red Tunnels, which linked millions of underground cities. There was a famous saying among Jupiterians, "Whoever controls the main roads will manage their intersections," and as the old Jupies remembered, "A saying becomes a habit and the habit becomes a law and the law stays in force as long as it's not broken and the breaking of the law is what the Evil Lords do."

If it was not just a planet, it could be called Hell; for the folk were only spirits carrying tarnished chunks of clothes or body parts, some looked alike, among smokeless but raging fires, and among constant wailing or deadly silence. The bodies were translucent brown and black, often compared to amber, in which old, well preserved parts were stuck for ages but these parts, loosely winding and disintegrating, were not well preserved. Yet life flourished; the folk were fighting, bellowing, running, jumping, weaponry building, but never sleeping. The Red Tunnels connected the cities. Armies moved through the tunnels from one battle to another, slaying anyone on the way. And battle it was that Jupiter was about; a senseless killing and humiliating. Everywhere the common words were like, (if one could hear these in the tumult of other horrifying sounds),

"Attack!"
"Smash!"
"Destroy!"
"Retreat!"
"Death!"
"Fire!"
"Kill!"

There a handsome ghost stared at a gold and yellow flame and talked to it. His name was Donkey.

"So who are you?" asked Donkey.

"I'm a tiny demon living in flame," said Blazing Night.

"Yeah, that I can see, but how come a charcoal smoke comes out of you?" asked Donkey.

"This is how I produce my energy," replied Blazing Night.

"So how can I fight you? How can I kill you?" asked Donkey. "How do I make you suffer?"

"Well, Donkey, you don't fight me, you serve me, do what I tell you!" bellowed Blazing Night.

"Blah, blah, I listen only to the Jupies, our Evil Lords," retorted Donkey.

"I came to Jupiter to kill them all, to trash the Evil Lords and to rule instead!" asserted Blazing Night and her flame went orange for a second.

"What!" said Donkey. "How could this be? You're only a tiny candle?"

"You're going to help me or I burn you to dust, now!" shouted Blazing Night.

Donkey was not easily frightened, after all he was a brave warrior, and in favor among the Jupies, but this time the seaweed smell and the white smoke made him shiver in a different way, not like piercing heart or eye, which was the usual way, but more like drying all your internals. "Sure, I can try this game," he thought.

"Let's do it!" said Donkey. "Let's kill the Evil Lords!"

Donkey quickly turned around and attacked a Jupie, who was watching him from a Red Tunnel. Blazing Night stayed still.

"Boomack, I'll smash you now!" yelled Donkey.

What could he do to a lord who ruled over seven Red Tunnels, which connected the city of Doomby to its seven neighbors: Mingo, Barek, Filth, Ham, Been, Eamer and Zooma. All citizens of Doomby were entirely enslaved to Boomack, but the ghosts of the nearby cities paid tribute to him as they had to other Evil Lords as their primary masters.

Boomack did not even move the tip of his nose; his bodyguards squeezed Donkey into pie, slashed him to pieces and kicked his parts into the fire where Blazing Night was. Donkey felt pain and only pain. As in every battle his roasted parts connected back. Another pain drilled through his heart. Donkey did not cry but squirmed like a dying dog.

Then slowly Blazing Night charged; first freed Donkey from the fire and quickly melted the two body guards in such way that only hot sand was left of them.

Boomback looked and waited until his guards could revamp and go back to the fight. But they did not go. Once a ghost was melted by Blazing Night, it did not recover, only charcoal and white smoke lingered in the remains. And the Jupie could not change that.

Now Donkey looked at Blazing Night with more fear in his eyes than ever. And his parts trembled and tears fell.

"Cool off, Donkey!" said Blazing Night. "Now you Boomback, pay homage to me like Donkey does so I won't melt you!"

Boomback was scared, "OK, what do I need to do?" he begged.

"Array all your slaves in Doomby and attack Mingo, I'll follow you and burn to sand all who resist! You're the Governor of Jupiter now."

So they did. Immediately the crowds marched to Mingo. Most spirits became the slaves of Boombacks but those who resisted were burned to sand by Blazing Night.

"Go to win!" yelled Blazing Night.

And they did. As usual the cities were burning. No hope, only despair was present, yet in some a charcoal-like smoke lingered over sandy remains of those who opposed Blazing Night. Perhaps more than one billion of such sandy stamps were there in only six days since Blazing Night came. Jupies were conquered. Yet some hid themselves in the deepest and driest tunnels where Blazing Night did not go.

The time went on and Jupiterians managed to learn the new way. Jupies stayed in the deep, well hidden dungeons and there every day marshaled their spirits to go up and fight, and kill all in the higher cities where Blazing Night ruled. Closer to the surface Boomback arrayed his troops and these fought back with rigor. Screaming and crying did not change things. The pain continued to drill everybody. Only once in a while a deadly silence prevailed for a moment between the battles, just long enough to make the spirits remember in what hurt they were in -- otherwise they could forget the pain when engaged in war. There was no hope; only despair.

Yet year after year, Blazing Night brought new equipment to Boomby and its nearest cities, built factories and laboratories making the spirits work with robots. Donkey was in charge of these efforts. His body was mutilated; complex mechanics and circuitry were added and stuck loosely among his translucent parts. He became a ghostly robot. He thought, "I might get a second chance and reduce the daily horror". And that was what happened when working for Blazing Night. Instead of additional hurt, a bitter, tedious will to push things forward dominated him. All had to be done quickly, effectively and instead of pain, a constant rush and feeling of urgency overtook him completely. His only thoughts were, "How could we speed the tasks? How could we do it better? What would work in a more efficient way?"

Now Donkey was a stubborn fellow; he might had learned that from the Jupies as he considered himself a close friend of Boomback and breaking the law was always in his mind, yet he was scared to do it. Being in charge of all that work for Blazing Night seamed to be a perfect opportunity to try what only the Evil Lords did. Then he found the courage and talked to Boomback.

"Boomback, how do I break the law?" he asked.

"Who are you?" demanded Boomback. "Go back to war! You're lazy junk!"

"I'm in charge of Blazing Night's work in Boomby, Mingo and the other cities," said Donkey. "My name is Donkey and I am the right hand of Blazing Night!"

"Oops," said Boomback. "So what do you want to know? A high servant of Blazing Night is like my friend."

"Yes," said Donkey. "My name is Donkey."

"Ok, monkey what is that you want? To break a law?" asked Boomback. "What law?"

"I don't know. That's why I asked you …" said Donkey.

"Ok, I'll do a favor to a monkey friend, go and scatter the smoking sand of those who were burned," suggested Boomback.

And so Donkey did for a few. Then he again felt pain only and could not focus on efficiency, hid himself in a dungeon in Boomby and waited in horrible pain, a deadly silence surrounded him and the waiting hurt all parts, even the circuitry overheated and almost melted. Then Blazing Night came and spoke from a tiny flame.

"You, Donkey will build a prison and will be the first inmate," said Blazing Night. "Don't show your face until you've done the construction and stay in the prison for one hundred years," she said and left leaving behind the salty, seaweed scent and a crimson shade, which soon disappeared.

This was the end of Donkey's career in High Service, from now on he was, again, only a spirit to fight and endure pain and wait for it in prison for one hundred years. Yet the circuitry somehow survived and several parts restored themselves to function. Something was telling him to escape and that thought only raised his pain to such a level that he could think clearly only a few minutes every day, during which he planned his flee from his prison cell. It was a tedious task yet the mechanics and circuitry were perfect to deal with the repetitive work. Slowly he figured out that Jupiter was a planet and there were other planets but he did not know how many and what their names were.

"Tomorrow is another day, perhaps I'll learn more," Donkey repeated to himself in his cell. "Perhaps Blazing Night will be merciful and let me go."

The works of Blazing Night continued. Her new High Servant was now an Evil Lord, whom she found in one of the dungeons. Her name was Bunter. Boomback and Bunter were partners, he was in war; she was in construction. He fought battles every day; she built things. Jupiter was as before, war was everywhere but close to the surface industries were born in many cities; factories, laboratories and prisons grew in Boomby, Mingo, Barek, Filth, Ham, Been, Eamer and Zooma. Over the years more cities were converted to that matter, yet the deep dungeons and tunnels were governed by Jupies, and Blazing Night did not go there, perhaps because it was too hot and too dry, or perhaps she did not need more space to build her elements on this planet.

She made many to work for her. Her main labs, factories, prisons and storages were located in the misty land of Jupiter. Humans did not wander there.

Nevertheless, she did not yet find her family. The closest she got were four candidates: Bruce, Adelaide, Frank and Joe: the four emerged to be very close to the genealogical inheritance she had identified as hers. Bruce's grandmother was Ann, Ann was a descendant of Christopher, she was the twelfth generation after him; Adelaide was a direct offspring of Rebecca, there were fourteen generations between the two, Rebecca was the only sister of Rob, also one of her ancestors was Diana, the great-granddaughter of Sanae; Joe's mother was Mirta, Mirta was in the direct line, eighteenth successor of Sanae; Frank's mother was Jennifer, Jennifer was the nineteenth generation of Gloria, also Frank's father was the seventeenth descendant of Gladys. Perhaps this track led Blazing Night to her family at last.

Chapter 4: Yellow Car

Monday morning came sunny and mild. It was late October but did not yet feel like autumn. The leaves were still green and cheerful. A refreshing west coast breeze carried a few clouds toward the main door of the car shop by the ocean's rocky beach on the island. The gold, orange and yellow beam of fire spiked steadily beside a pile of pastel pebbles. Blazing Night was present. Her soothing, white, charcoal incense quivered in the air. Her seaweed scent blended well. She was watching and listening, yet hiding among the rotten wood, filthy sand, dead crabs and pale stones.

There stood the four players who hoped to triumph in a game of which they knew nothing. Days ago Frank had lost track of her; he had led the run in a dim space but instead of winning had crashed his car on a left turn, and landed in the Dark and Warm Fluid. This proved disastrous. Junk and scrap was all that was left of his equipment. The four dragged it here, placed it in front of the broken door among oil, tools, heavy parts, gizmos and now learned that repairs were pointless.

"All right, all right, get a cold drink!" Bruce huffed nudging his big ruddy nose forwards a bit. "Are we giving up or what?"

"Blah, blah, blah, Bruce, why don't you help us… make a gadget or something?" asked Frank and shook his ash blond head.

"We've gotta win!" shouted Joe, tapping his fox red hair.

"Calm down," snapped Adelaide and stirred her deep brown eyes, which generated a glint of white light. "Like we need to fight while the game's on and we can't get to her… yet!"

At last they stopped talking. Their clones pulled in, reported progress, waited and listened to new instructions, learning how to act.

Still, the four ignored the new comers. Frank shifted his right hand in front of his face as though pushing away invisible flies. After a moment of frustrations Bruce briskly shelved the four doubles away.

"What dummies our clones can be! Can't they be more independent and know what to do ... rather than us telling them ... and wasting our time?" asked Frank. "They're a bunch of lazy morons!"

They often irritated him by being dull, boring and uncreative. He had to explain every detail to them for they required lucid and precise commands, orders, and directions just to operate and serve their owners.

"Frank, grab your blazer and buy a yellow speedy car!" replied Bruce closing his fists. "We'll search ... and find her... all of us, we take one new car ... and Joe's driving."

"Do it!" responded Adelaide quickly.

"OK, OK, I'll buy it ... now," muttered Frank and did as he was told.

In a few minutes, just in front of them, stoop a car and a ship. Its oval body glittered yellow light. Jets and pillow wheels carried its stomach. Three hundred bars of gold was its price.

"Get aboard!" yelled Joe.

He did not have to say that twice. All of them jumped at it. Yellow Car appeared spectacular and felt comfortable once they sat inside. They took off. The road was broad and let Joe hit the speed of five million knots in few seconds. Bruce was watching for any sign of that dreadfully Dark and Warm Fluid.

"Look at the green light on the screen, just where we're heading," said Frank smiling. "That's where a creature attacked me twice."

Suddenly two quick shots were fired. The vehicle got hit in its left side not that far from its engine. Adelaide flipped from her seat to the trunk. Bruce seized his short blazer and pressed the pistol. Reddish laser beams and blue jets of fire were gushed from the underbelly of their yellow machine. Bruce held tight to the railing of his chair.

"Got it!" said Joe, widely opening his bronze eyes.

The car did well. Its body had not even been scratched. Its passengers appeared to be satisfied with its performance, showing their teeth. Bruce dried the moisture of his palms against his pens. Adelaide smiled.

"Expect more monsters?" asked Frank and pinched his sideway-bended nose.

"We'll do all right," whispered Adelaide, her black hair stuck flatly to her head between her small, round ears.

Little pieces of the remaining creature's fixture, its weapons or its armor were flying in front of their capsule and could damage its insides. Joe slowed down to one million. They saw greenish, reddish and greasy particles, which certainly could reassemble if the right rule was available and one would know how to run it.

"Joe, speed up!" Bruce urged impatiently, moving his big head forward. "The debris' center must be behind us and we need to drive faster."

It seemed like a second, yet the crew reached ten million knots, where slow or less powerful creatures could not catch or notice them at such pace. Some gaming experts called that a safe speed.

"What's that?" asked Frank pointing to tiny two dots on the map.

"It's the same dark and warm fluid you fell into," answered Adelaide, showing her sparkling white teeth. "I'll bet we're surrounded."

About twenty large patches of dark, messy droppings instantly covered half of the car. The entire console turned black. They held their blazers, shivering.

"That's not the same fluid," whispered Frank. "We're in a parking lot."

Joe stopped the car. Their clones had just checked in after school. In the next few minutes the players explained what to say and to do at home. Frank's clone was the smartest and the most expensive one, and could alone invent a simple rule that painted a fence or cut grass while others needed regular upgrades just to do what they were told to. The originals shouted a few words to their doubles and sent them away.

"Frank, you'll drive now," said Bruce. "We've got to get out from here, fast!"

Again the engines roared and hit four million in two minutes. Yet the track was different; stuck with masses of trees and no road signs, and its surface seemed to be substantially darker. Adelaide almost sunk into a fanciful dream. Her head did not move but her eyes remained open.

A bright blue star was following them like a shadow.

"Shoot it!" shouted Joe. "What's that?"

"It's a spying star," said Adelaide waking up. "We're being watched."

It boomed! A loud explosion centered just in front of them; red and yellow blasts everywhere. Immediately two brown, dusty monsters ran behind and were ready to kick.

"Faster! Faster!" Joe shouted. "Get to twenty million knots, thirty millions, and forty."

The monsters could not fly any quicker. Their yellow vehicle was exhausted too; the engines were vibrating loudly. Its yellow body brightened; gained a new, orange shade.

Then Adelaide slowly took a portable mine from her pocket, set the sensor, and let it go by the rear window. A dark cloud rolled behind and quickly vanished. All monitors cleared.

"Slow down," shouted Bruce as though he worried about the car or the monsters, no one really knew which. His face was wet.

Several moments later Frank halted the speedy new machine beside a bunker-like hotel where they lodged the one night. They all understood well that tomorrow was also to be a day full of battles and bruises, and soon they fell dreaming into a deep sleep.

"Good night," whispered Yellow Car.

"Quiet!" said Blazing Night shaking her crimson shade beside the car.

Yellow Car, shivering in the parking lot under a maple tree, looked at her gold flame but did not move or say anything. She burned for one hour and left living her soothing scent behind. Only then the car engines cooled off.

Night came softly and sweetened the dreams. Then the wind brought a chill. At the beginning of the morning, the four woke up in their rooms, and stepped down to the lobby.

"No one is ready to play," said Joe. "We were working hard last night."

"Is that so?" said Adelaide. "Is eight hours not enough for a fine rest?'

"This place sucks… it is like a tiny bed and skinny breakfast station," said Frank. "The cook is vacationing; the lazy keeper serves coffee and scrambled eggs every meal."

"Hi," said the keeper to Bruce. "It's time to pay the bill and drive away."

"We'll do that," replied Bruce quietly.

After paying their dues, they left the building and met with the clones. Bruce was eighteen years old, Adelaide twenty two, Frank fifteen and Joe had his twelve birthday just one month ago. Their doubles looked alike but they smelled different; the scent of decay followed every clone, yet only the originals and their friends could sense it.

"Remember this," said Frank to his clone. "Don't ask my teacher any questions in the morning, just smile, give her an hour or two to warm up. My marks must be tops! The math rule we're studying seems to be a complicated one and we have no clue how to formalize it yet. Make clear to her that you're working hard on it and you'll provide a very sound solution within the schedule that was so precisely explained by her. Make notes continuously and hide yourself behind the back of others. Don't let her question you! Whenever you see her willing to interrupt your constant quiet attention swiftly move your lips and your body like you're perfectly prepared. In case you're selected tell her you're not exactly well and haven't seen a doctor yet. Otherwise act as you did in the most recent day you recall," Frank concluded.

Then the clones went to school.

"Why don't we look for a restaurant to get some food to kill the taste of the coffee and sandwiches from the hotel?" Frank suggested.

"OK, OK, let's go to that neat looking place on the corner, just in front of the lobby," Adelaide agreed.

So they did. There they sat down and ate. The waiters were very nice.

"The breakfast is great. I feel rather heavy now," said Joe and stared at the window, scanned their bunker-shaped dormitory, perhaps he wanted to return and to dream there.

"Yep," said Bruce. "We must go. Frank, you will drive from now on."

They crawled into Yellow Car. Frank started the engines. The sound of it was cheerful.

"What a highway it is!" said Bruce. "It has thirty lanes and no traffic on it."

Nothing interesting could come out, it seemed to him.

Then the Dark and Warm Fluid covered three sections.

"We're on the right track!" shouted Joe. "We're back in the game! Hey, what the …!"

A sudden earthquake cut the road in two. A vicious monster came from nowhere and kicked the car sideways. It was a fast, small creature but had done enough damage to cause a few bruises and change the path they were taking toward a mountain, enough to kick them out of the game. They came so close.

Yellow Car was dead. The players ran. Their dirty faces were scratched and wet. They smelled bad. Even clones did not stink so.

"Where are you Bruce?" shouted Joe.

"Adelaide, help me!" cried Frank.

"Where is Bruce?" Adelaide called out. "Frank, pass me your blazer! Joe, see my big, black gun? Take it and start shooting!" she ordered.

Under heavy fire Adelaide was creeping toward the wreck of their only vehicle. Bruce was trapped motionless between the wheel and the seat. She pulled him out and gave him his favorite cold drink. He could not say anything. Just took a pen and wrote the Yellow Rule on his blazer then connected it to the broken monitor of the car.

"Hope it works!" mumbled Bruce, spitting saliva and blood.

One by one every yellow part changed color to gray and produced a black cloud so dense no weapon sensor could penetrate it. The parts collectively flew up north and joined the engines to compose a new or rebuilt stronger Yellow Car which at first targeted the monster while its seats jumped out from its body and loaded four players on board. In two seconds the car hit sixteen million knots, turned left and sent a small rocket toward the surprised monster.

Frank took over control of the vehicle and drove it far from the blood and stench of the battlefield. But that was not the end of it.

"Look … three monsters in front of us!" shouted Joe.

Bruce ran to the left window and immediately opened fire from the door. The number of bright, white lightning flashes touched the enemies just enough to freeze them instantly. Lucid pillars of ice rested behind. Bruce laughed aloud and shook his shoulders.

Yellow Car stopped near their emerald painted school beside which the clones were ready to report their progress and ready to get instructions for the upcoming evening. Their faces looked stupid. They opened their mouths, nostrils and ears gaping at the originals.

"We bumped into few difficulties... today," said Bruce to the clones. "Use the repeat last evening procedure."

So they did, and went home sighting their originals in a discrete way as though they needed more instructions. But there were none.

The players reviewed their plan of the newly developed defense strategy; mounted two blue and black bikes to the rear of Yellow Car just in case when the car broke and failed to function just when a player needed to counterattack rapidly and decisively. Then they climbed the vehicle and drove away.

"Joe, where are you going now?" asked Frank.

"Straight forward and fast," Joe replied.

Within a few minutes the car hit twenty million knots. Suddenly a big and nasty creature appeared from the south. Bruce warned it that they were fearless players and it had better ran or died. It was a Brown Monster and it decided to cooperate and gave Bruce additional information about the game of which they were dreaming.

"The rule to find it was unknown, but the name of the game could be," the monster announced.

Under these circumstances Bruce let the creature go free. He could not stop thinking that they learned the way; they had to discover the name first and he was convinced that it was the key to the first door of the challenge. This was the step of achieving the advantage so needed in the battles, this was the weapon used in inventing a winning rule in the game and who did not want to win and who would not do all including choosing a painful and a dangerous, long road to victory. If Bruce were a clone such thoughts could burn the circuits. But he was not. He continued thinking. Everybody wanted to play this game but only a few were chosen and were capable of winning. That night he thought that they were led to an adventure of challenges not seen or heard before in their universe of games and schools. Perhaps none of the other fellow students might even comprehend what major progress they made Tuesday. He could not sleep thinking and imagining the new path they were on and what it might bring Wednesday. There was glory or failure in the making.

Seven and half hours later the four duplicates found that their originals were still in bed, walked up to their rooms, knocked on the doors, and spoke to them gently.

"Hello, Frank!" said his clone, "It's time to go to school."

"OK, OK," mumbled Frank.

A gorgeous morning, warm sunshine and a refreshing ocean breeze encouraged players to start the day, but late night dreams prolonged the restful sleep. None got up.

The clones gave up and went to school without listening to the usual commands and clarifications. Their confident steps indicated that they could survive one day without any special instructions.

And Adelaide was first to enter Yellow Car and to check the equipment. Several gadgets were new and appeared well fit to challenge extensive, dangerous trips.

"How are we going to learn that name?" asked Joe.

"In a battle!" answered Frank.

"Not really," Adelaide said. "It may be a race or a treasure hunt where we use our best rules."

"It sounds familiar," Bruce agreed.

"When this will happen?" asked Frank.

"That's a good question," replied Adelaide.

"We may have to study new rules and get our clones to explain them ... based on the details they remember from school," Bruce suggested.

"My clone says that the last rule he heard about might help in finding the name of an unknown," Frank informed Adelaide.

"Sounds great," Adelaide replied.

"Can we get some food soon?" asked Joe.

"I'll order a burger, a salad and an ice cream ... if you agree," said Bruce. "We'll eat in the car."

"I want vanilla and strawberry please," said Joe.

"What are we waiting for?" asked Adelaide. "Let's go!"

"Oh, yes," they all said.

After loading the car with tasty food, Bruce drove carefully at the speed of fifteen million knots. Soon the players finished licking the ice cream. That day their highway was old with several holes which had not been repaired for months. Bruce had to follow not only the thin path between the holes, but also had to check for what was in it. Adelaide watched the road behind. Their blazers did not indicate any unusual condition. Frank and Joe took a nap. Then the car stopped.

"Hello, Frank," greeted his clone, opening the door.

School was over. The clones were ready to report and to explain all new material they had studied. They gently approached the originals.

"OK, OK," Frank grumbled, getting out of his comfortable sit, "Clarify the meaning of the rule we're studying at school. Go ahead!"

"This rule is an axiom and deals with multiple views of reality and tells us how to seek them intelligently and perhaps find a hidden name or a clue. The color of the rule is beige. It was discovered a few months ago. Once the axiom is established, the entire search could be more efficient and resilient, thus it could be conducted in battle. An example of Beige Rule use is the selection of the winning weapon in a duel. Still lots of research to be done before it can be deployed in the field. Students are involved in doing prototype testing. A friend of yours, Paul, has already found out that when testing it in the class, the beige axiom once connected to a quiz wrote all the answers correctly and quickly. The teacher also presented a case where children were too loud and the Beige Rule connected to the class told them to shut up," said Frank's clone.

"The Beige Rule is kind of boring, isn't it?" Joe thought out loud.

"But we need it," Adelaide pointed out.

"We're going to test it tonight," Bruce ordered.

Then the players sent clones home and drove to a near, sandy desert. Frank and Joe mounted the two bikes. Bruce prepared a prototype of the Beige Rule. Adelaide controlled Yellow Car. They were busy.

Suddenly a cute looking dragon came out of the sand and fired at Joe. Joe ran away. Immediately Bruce connected the rule to the desert. The dragon was stupefied. It could not move or fire. His eyes popped. Frank inspected the creature. His skin had spots as though he had burned himself. Yet the huge eyes and the big mouth resembled someone pleasant or familiar. He touched his taupe nostrils and huge, egg-like eyeballs. It was one of the Beige Dragons who live in barren and scorching regions.

"It's harmless," Joe reported.

"What is the name of the game?" Adelaide asked and raised her hand as though she wanted to press the trigger of a pistol.

Instead of any response, a sand storm kicked in, impacted everyone in the car, and relocated them to the nearby hotel where they had slept two days ago.

"That's a side effect of the Beige Rule," informed Adelaide.

"It needs more testing," noted Bruce.

"What did happen to the dragon?" asked Joe.

"He's buried in the desert," replied Frank.

They went to the hotel and did not think about anything else but the beige sand. Adelaide slept in a big, round orange room. Bruce was in a dark blue square one. Joe and Frank shared a wiggly, cozy, green and yellow space.

"Good night," whispered the creature, grasping sand out of his way, but the players were in their beds far away from the desert. Only a little crimson and gold blaze gaped and breathed by the bush behind the neck of the dragon.

The morning of the next day, Thursday, was not peaceful at all. They were sitting in Yellow Car sweating and waiting anxiously. If it were not for climate control, the stench would make it impossible to stay calm. It was the first time the vehicle was due to transform into a spacecraft. Its yellow bumpers reshaped to jet engines. The capsule rounded the oval-like body on the top. The interior stretched and the controlling panels allowed accessing the rows of new, shiny buttons.

"Are we ready?" asked Frank.

"I'm good," replied Adelaide.

"What are we waiting for? Start!" Joe barked.

A huge cloud of smoke covered the landscape of the small town. All red and blue engines were turned on. Yellow Car became a rocket and lifted off into space.

This trip was necessary to investigate and possibly challenge a crowd of Brown Dragons up in the space between Mars and Blue Earth. About five hundred creatures of that kind were reported to have gathered there to meet another group of three hundred Brown Monsters. It was not known why they met, yet such a number of them was a threat to the players. Usual activities were upset and those who were assigned to this mission were leaving their towns without delay at 6 am universal time. The command of the operation was given to Officer Brenda Hawkins. Adelaide, Bruce, Frank and Joe were part of that fifty-ship team. All quickly arrived to the destination and assembled, ready to join in battle.

"It looks like an invasion," Adelaide asserted.

"Prepare all weapons," commanded Bruce.

Frank drove Yellow Car, now a spacecraft, toward ten monsters.

"Attack!" shouted Brenda.

Hundreds of brown creatures gasped at this splendid and colorful army. From tiny spacecrafts to large spaceships of different shapes and sizes all had fired in one instance. The explosions of red, white and yellow rocket bombs and laser beams astounded everyone including the attackers. Like tremendous fireworks, the dark space became a place of battle and this was only the first shot!

"Attack again!" shouted Brenda.

Half of the Brown Dragons were running away, half were defending themselves; a hundred returned the fire immediately, another hundred flew aside and attacked this extraordinary army on the left side, which appeared to be weaker for a moment.

Five monsters jumped on the top of a blue spaceship and kept kicking and punching it to the despair of its crew. Cracking sounds could be heard and could mean only one think: the end of the craft.

"Frank, turn and help the blue ship!" shouted Bruce.

"Joe, start shooting!" yelled Adelaide.

Joe liked this kind of battle. Everything depended on a fast, exact fraction of a second maneuver and only one precise shot. The need for accuracy made Joe shiver. And then his skill was displayed by his almost perfect control of man and machine; his face white, his fingers calm and the decision just on time to press the trigger.

"You got one!" shouted Joe.

Yes, he shot it, yet three more dragons flew from behind, caught their yellow machine and held it so tidily that it lost speed, almost stopped and became an easy target for anyone to trash.

"Retreat!" commanded Brenda Hawkins.

But it was not possible for the crew of Yellow Car to move. Large monsters, now twenty, held the vehicle and were close to dismantling it within few seconds.

"Any idea how to get out from here?" asked Frank.

"I'll do the Beige Rule again," replied Adelaide.

She connected the new prototype to the car, and typed the Beige Rule on her blazer. The monitors showed a Beige Dragon approaching that soon drove the monsters of the yellow ship to a dark space far enough distance to regain the control of the flight and to retreat to a safer corner beside the splendid army of space machinery from Blue Earth.

"These monsters are mightier than we are," admitted Brenda Hawkins. "We attack again or perish if those who ran away come back and fight."

An entire army of fifty ships flew swiftly toward the cloud where the battle was expected to be a feverish struggle.

"On my command, fire!" shouted Brenda.

They could not see anything, but were shooting at the cloud and directing the great ships towards the center of the smoke, hunting the enemies. But there was no trace of the Brown Dragons or Brown Monsters. The hunters exited the cloud, checked the sensors, and saw no indication of anything but the brown creatures that ran away before and even them being too far to pursue.

"Mission accomplished," shouted Brenda, shivering. "All crafts return to Blue Earth!"

Back home no explanation was found as to why and how the monsters and dragons had disappeared in a cloud in the middle of the battle. However, Brenda Hawkins received a silver medal for bravery in action.

Blazing Night knew all the details. It was clear to her that that fool, the Beige Dragon, had knocked out several of her finest creatures, which were in progress of regrouping. It was her plan to consolidate her forces and prepare them for a major battle.

"He'll pay for it," she said to herself. "I'll find him one day."

Hours later the four slept soundly, then woke up rested, and began to dig data. It was Friday. Life was as usual; plenty of appealing facts waited to be gathered and analyzed.

"I like it!" said Adelaide. "Frank, buy this book."

It was one of those things they did when browsing in a warehouse of used and forgotten works. That rainy morning players stayed in hotels, and later looked for items of interest wherever such were advertised and available. No doubt Adelaide was an expert in finding exciting and brilliant stories to read or to learn from. Some tales contained clues to use later as items like keys to unopened doors.

"Where are our clones?" asked Joe.

"They're away getting new upgrades," said Bruce. "It's a battle against a virus which can read their memory and often cause damage to their neural system. Even Frank's clone had to go. It's a severe alert."

"Who's going to school for us?" asked Frank.

"We are," replied Adelaide.

"Can we skip it?" asked Joe.

"No, no we can't," answered Adelaide. "We are back and we will study again."

And they marched to the school. It was morning, the bright sun bathed in lazy clouds. The four players entered the school gate together and each went to their respective classes: Joe in grade two, Frank in grade five, Adelaide in eleven and Bruce in nine. They did not talk to each other in order not to expose their team to students or teachers. There were players in the school who could take advantage in a game once they learned the names and the organization of the team. It was already high risk having all four of them to attend one school.

The school was called St. Albert. Albert was a kind of hero and saint in olden days and some students learned his story from one of those old and forgotten books, but it was unknown why his name was chosen or by whom.

Frank was writing a math quiz that day. He had to work a few questions and to select multiple-choice answers. At the end he had to describe the Beige Rule, and this last task was considered to be his prime assignment for a few weeks. He thought about the fight yesterday and about the side effects they experienced and he also remembered a Beige Dragon wishing him good night. He came to the conclusion that something that was not explained had to do with the

Beige Rule, the dragon, and how the events surprised everyone. He decided to write about the Beige Rule as a great potential discovery already causing disturbance in the universe and he wrote that the rule was comprised of actions desired and actions unexpected.

"How did you write the test?" asked Paul during the lunch break.

"I did the Beige Rule section the way I saw it," answered Frank.

"That won't be enough to get a top mark this term," said Paul, laughing. "I'll win! You'll loose."

Frank did not know what to say but walked away and finished his lunch. Later he played soccer with other boys and girls. He scored a goal but his team lost anyway.

"Frank gets top marks for his story," said his teacher. "No doubt children, you are making progress and it shows in your work."

The teacher's name was Natalie Foster. She was an experienced and proud mentor. Frank liked her. She seemed a fair professor and despite the fact that he did not see her everyday he actually enjoyed his clone's reports describing what Natalie did or did not do. Otherwise the reports were boring, yet he had to listen to it carefully to keep himself learning.

Joe had a tough day. His teacher, Mary Rooster, asked him too many questions that he could not possibly answer. His mind was on the battle with the Brown Dragons and Monsters and how Brenda Hawkins commanded so well but did not get the gold medal.

Adelaide and Bruce walked out of school early. The work was boring. They felt sick and decided to rest at home.

After school was over the four players met beside Yellow Car, boarded it and drove far away from classes and homes to a calm and silent spot on a beach beside rocks and rotten woods under a huge root. Their clones arrived after the upgrades. Bruce sent them home.

All of a sudden the Beige Dragon came out and looked at them in confusion, not sure if allowed to proceed and talk to them or to stay in a distance and wait for an invitation. Several moments later Adelaide moved her hand inviting the monster to come closer and speak.

"How are you?" the dragon asked. "Do you remember me? Do you realize who am I?"

"No, we've got no clue what you are. Explain, be quick!" replied Adelaide.

"I'm the Beige Dragon enhanced by the Beige Rule and my duty is to protect you and help you to seek and find the name of the game and to help to win with her," explained the monster.

"What's your name?" asked Joe.

"Florien Beige," replied the dragon.

"Why are you helping us?" asked Adelaide.

"My master is Bruce, who created me from an ordinary monster when he applied the Beige Rule in the desert," Florien Beige informed her.

"What are your powers?" asked Frank.

"They are many, including ability to fight thousands of monsters and ability to disappear and to appear at places of my choice," said Florien.

"Where do you live?" asked Bruce.

"Master, the desert is my home and I dwell in the sand," said Florien bowing to Bruce.

"Do you attack other players?" asked Adelaide.

"It depends, if my master Bruce commands so I will," answered Florien Beige.

It was late, the players drove to the hotel where they drank a cup of tea and went to bed. It had been a long day, and they gained additional powers, surprisingly from the enemy camp -- the vicious monsters.

"Is it possible to transform a dragon into a helper?" Bruce said to himself before falling asleep.

The Land of Games lay calm.

Welcome Bunker was the hotel name. The four players lodged there for one night only, but kept returning again and again to stay and to rest. There was music playing in every room, a unique composition matching the mood and personality of the listener, yet not interrupting conversations, which continued among them.

"Where's Bruce?" asked Adelaide.

"I don't know," answered Joe.

"Where is Florien, the dragon?" asked Frank.

"He should be protecting us," answered Joe.

They jumped quickly into Yellow Car and drove off into the desert. It was a clear day and their speed was fifteen million knots yet they saw neither trace nor sign of Bruce. Their faces looked worried.

Then Adelaide prepared a prototype of the Beige Rule and connected it to the desert. A huge screen came leisurely out of the sand and a scene of Bruce and Florien Beige fighting a crowd of Black Dragons was displayed shortly. There was no time to delay a rescue operation. Adelaide told two boys to prepare for battle.

"Load your guns!" she shouted. "Frank, you're driving. Speed up to twenty million knots and fast."

"Yes, Sir," said Frank.

"Guns are on," said Joe.

"Joe, you guard the rear and shoot anything that follows us," said Adelaide.

"Yes, Sir," said Joe.

Within a minute they reached a group of monsters running toward the fight. Adelaide took a big gun and instantly sent a series of bombs, which hit the monsters and flipped them over several times until they dropped and sunk into the sand in such a way that only their tails stuck out.

"I see him, I see Bruce!" Frank was shouting.

"Attack the Black Monsters!" commanded Adelaide.

Immediately Frank started shooting so intensely that he did not notice what was going on around him. Joe was furiously discharging mines at any living creature which came near and was detected by motion sensors. One monster came close to a window, ready to hit and kick players. Joe exploded him with one portable bomb. Colorful stars flew at the monsters, disoriented them and caused them to fall or to become easy targets. A huge orange light was frequently seen among dragons cut their weapons and disabled their powers. This was a result of Frank's laser guns continuously and noisily punching like the artillery of an infantry battalion.

Hours of fighting did not slow anyone. The cunning dragons appeared confident to prevail and had gained an advantage close to winning. But the battle was not over.

It was their yellow vehicle that turned around and kicked five Black Dragons, smashed them and killed them. Adelaide killed five. Florien finished fifty. Bruce was freed. Players escaped unharmed. Florien Beige suffered a few bruises on his sturdy back.

"I was captured by the Black Monsters last night," explained Bruce. "Florien defended me and he fought bravely for me."

"There are about a hundred of them," continued Bruce. "They come from different games and communicate across the universe, calling each other to join the action. Thank you, you arrived so fast to the rescue."

Bruce was exhausted. It was no fun. They drove back to Welcome Bunker, very happy to be there together and to realize how frightening it was to be separated from each other, surrounded by enemies in the Land of Games.

"My name is Yellow Car," said the car. "I come from the game you're eager to play."

"What's her name?" Adelaide demanded.

"Blazing Night," whispered Yellow Car changing its color to orange for a second.

Sundays they did not play together. Adelaide and Frank attended High Mass in the Cathedral in Vancouver. Bruce studied in his room. Joe enjoyed games with his brother in the family home.

That Sunday the rector gave a long sermon and the faithful waited impatiently for a conclusion.

"When will it end?" Frank asked.

"Quiet!" said his father, annoyed.

Adelaide sat in the middle of the church. Frank was in the last row. The Sunday service was at 11 am local time; not many people attended, mostly elderly and weird looking grown-ups went to mass. The Clergy frequently complained about the wealthy panhandlers, who got a way too much attention and interest; everybody wanted to pass by yet making sure not to disturb the hard working top executives. Some went to church just to have a chance to walk closer to the panhandlers of which some were the most famous citizens in the world. Everyone knew their names and dreamed of becoming one. Others thought about their offspring, relatives or acquaintances that could get the job in some distant future. There were plenty of stars with famous names engraved in the sidewalk on the block. It was known that the senior, most successful players might join the most revered profession of panhandling, but who else could get that desired job was a mystery of unprecedented complexity. Only faithful regulars went to church every Sunday. Adelaide and Frank and their families did.

The entire congregation sang a hymn,
"I will sing to my God a new song:
O Lord, thou art great and glorious,
Wonderful in strength, invincible,
Let all thy creatures serve thee,
For thou didst speak, and they were made."
- JUDITH 16:13,14

This church had a beautiful large organ, its golden pipes trimmed in blue. The finest organist on the West Coast was in attendance. He enjoyed playing familiar melodies before and after mass. Some people often stayed for several minutes just to listen, and clapped at the end. Adelaide liked that part too.

"How are you doing in school?" asked the rector at the door as they were leaving.

"I'm OK, my grades are high," replied Adelaide.

"Did you get any new toys?" asked the rector.

"Yes I did," replied Adelaide. "But I can't solve a new, serious problem with the present rules."

"Good luck!" said the rector. "Have a great day now."

Sunday was a day when players could not dispatch their doubles. Instead, clones went to the garage, where they studied educational material and helped to repair the devices and machinery, which were always required to function in battles and games.

About the same time, Bruce was planning activities for the following week. It was a tedious chore. He analyzed tremendous amounts of detail using modern tools and rules. He often spent an entire day to conceive possible scenarios of events and how he could respond to them. That day the subject was how the Brown Monsters would invade St. Albert, and how such a fight would involve the clones, and how the players and clones would be forced to disclose their duality and to come to light in front of the teachers and students.

"This should never happen," he muttered.

"What?" his mother asked.

"Disaster," replied Bruce. "No disaster is allowed in our school."

"You pray daily," said his mom. "It won't happen."

After analyzing it again, he somehow found the words of prayers more assuring than his scientific calculations, which he performed in order to estimate the probability of such tragedy in the school.

Joe played with his brother all day. They built bridges and raced cars over and over. They both had two thousand cars each, and the race went from one room to another around a table, a chair off to the bathroom and back to the kitchen and five times around a table and straight to the finish line. They repeated it hundreds of times and kept a log of results for every car and driver. The log records were presented to the parents with the explanation of why a particular driver or a car was faster and won.

"What color is the fastest mom?" Joe asked his mom before going to bed.

"Good night," replied mom.

It was past midnight. The sky was dark. The dispatch body of Blazing Night passed among rocks and sand on Second Beach beside Stanley Park. Her crimson shade flickered behind her. The scent of seaweed surrounded her bare, round shoulders. Now when they met, the body and water merged and the bright head and trunk swung and the ocean puffed with golden and yellow rings. Then the rain came.

65

Chapter 5: The Land of Games

The four gaped at the crimson and gold blaze warming their bodies. There was no shape to it, yet there was a smooth, shiny, vivid, breath-like presence.

"Are you ready to start?" asked Blazing Night from the fire.

"That's a good question," Adelaide agreed.

"Are you exhausted?" asked Blazing Night.

"We'll do all right," Joe interjected, smiling.

"Go for it!" Adelaide urged.

Yellow Car quickly dived into Blazing Night's dark fluid. They did not know where they were going to land once out of it. A tiny crimson and gold flame rested on the shore of the fluid, then disappeared as though it had never been there.

They popped up in a big city. There were a lot of vehicles flying among the shiny and glassy skyscrapers. There were no monsters or dragons around; it was a peaceful community of modern generation clones.

"Looks like a simple day ahead of us," said Bruce.

"Maybe not," cautioned Adelaide.

"I like it!" insisted Joe.

"I'll drive," said Frank.

"We'll all do our parts," Bruce added.

Bruce could not stop wondering how many times a quiet activity had turned into violent battle in a matter of two or three seconds. This was the way of Blazing Night, the game they were so eager to play.

She was like a viper fast changing directions to attack or to defend. Her crimson and gold flame was fabulous and famous. The smell of charcoal revealed her passing by. Her crimson shade pointed to a spot from which she had moved away. Her seaweed and salty odor could almost be touched if one were close to her. It could not be discovered in how many places she dwelt at the same time, yet one consciousness and one will made her whole. Some suspected that only when seeing her dispatch body did one see her entire being, and it was only then that she dwelt in one spot at the moment. Bruce continued dreaming and silently asking himself: was there any other game like her? Every day she was different, like a chameleon changing colors, so Blazing Night switched spaces and surroundings yet it was as one game. Bruce often asked how many players were seeking her. How many knew her name? How many played her regularly?

Four players knew and could enter her at any time they wished. They were Bruce, compared to a terrible crocodile, Adelaide, identified with the personality of a merciless anaconda, Joe compared to a swift scorpion, and Frank frequently viewed as a bold, relentless vulture. Again Bruce dreamed: who could face them? If that were not enough, they had two helpers: Yellow Car, their robust vehicle, and Florien Beige, their charming and wonderful dragon.

Cruising high around the structures, they spotted a little male fellow in a blue shirt among a multitude of city clones. He was waving toward Yellow Car.

"I'm driving to land," said Frank.

"Watch the blue guy," cautioned Adelaide. "He looks suspicious."

"We're slowing down," said Frank. "We'll park here."

"I'll get out first," said Bruce.

As soon as he stepped onto the balcony of a building where they had docked, a brown, smelly monster shot twice at him but missed, blinking its huge yellow eyes. Joe kicked it hard. The brown, dusty, bloody, rotten carcass fell down toward the crowd of clones and the vehicles soon disappeared.

"Can't even enter a building without a shot," Bruce complained.

"Welcome to Brown Dragon Kingdom," said the blue clone.

"Where is your king?" demanded Joe.

"In his palace," replied the blue clone.

"We've come to win," said Adelaide smiling.

Then the blue clone ran away. The players returned to the car and set its sensors to find the royal palace on the map. Hastily, Yellow Car figured out all the directions. Then they set to go there.

They arrived in ten minutes. It was the Spectacular Brown Castle, well guarded by hundreds of dragons and monsters. Clones were not permitted to come close to it but they could watch it from afar and see it only when the weather cooperated and the visibility was sufficiently clear.

"We'll attack through the front gate," said Adelaide. "Frank, faster... get it to twenty million knots!"

"Done," said Frank.

"Attack!" shouted Bruce.

This was not a typical battle. Four players were shooting at any moving creature. Yellow Car dodged about to avoid hits from their enemies and with the velocity of twenty million knots headed straight toward a little hole still opened between two pillars of the gate. Many explosions looked like a composition of light, not a battle. Red, blue and yellow colors mixed with smoke, cloud, noise and shouting.

They entered and saw the king, a small Brown Dragon with a gold crown and a long sharp tail set with flashy gemstones sticking out.

"We're here to win!" shouted Joe.

The king's bodyguard surrounded Yellow Car and ejected green grenades at it. Frank could not speed up to thirty millions but drove about ten million knots in a circle while Adelaide, Bruce and Joe were discharging portable bombs and mines. Brown, sweaty, stinky monsters were catching up and attacking from every direction. It was very close to a shameful defeat. Bruce wanted to command a retreat but stayed calm instead. Joe shot at the king and missed. About a thousand creatures helped the brown royal to hide behind the golden throne. The temperature grew too high to control the battle. Some monsters melted and turned into a boiling swamp. Adelaide triggered the freezing guns in order to abate the heat. The efforts of the four did not suffice to win and came too late to defend against the enemies. Huge crowds of dragons stuck to Yellow Car, dragged it to the ground and built a pile of bloodied bodies on top of it.

Then a big paw grabbed the vehicle and took off into the air. It was Florien Beige coming to the rescue. They escaped.

In a few seconds the players found themselves in Welcome Bunker. It was a warm evening. They were safe and sound, but tired and exhausted. Yellow Car had to undergo substantial upgrades and was transferred to the nearby repair shop. Once more the brave four fought treacherous battles and were left with bruises to heal. Bruce could not stop asking his brain: did we learn new tricks in order to thrive in the Land of Games?

On the other side of the street beside a lamp, a yellow and gold flame lived and watched.

"They're not that bad, are they?" thought Blazing Night. "Perhaps these are my relatives."

Obedience was something they had to learn. They had no choice. Every player had to respond swiftly in combat. Often there was no time to think, only to act. Many times Joe and Frank tried to respond exactly to the orders, but perhaps due to their inexperience they did the opposite. Instead of shooting the Black Monsters they shot several white creatures, which looked dangerous but were harmless. The game instructions said, "Do not shoot white!" This incident was reported to the police and the two boys were prohibited from playing one day a week for two months. It was today that they did not enter Blazing Night. They stayed in Welcome Bunker and kept studying reports prepared for them by their clones.

Adelaide and Bruce drove at a snail's pace into a small town. There were few empty houses. It was a settlement of ordinary clones, which mined gold in nearby mountains. Every hour a huge Gray Monster flew by.

"Speed up to twenty millions, Bruce," Adelaide ordered. "Let's shoot one."

"Sure," said Bruce. "We'll catch it in two minutes."

It was an easy mission. The Gray Monsters were huge and traveled alone. Adelaide took a short gun and loaded it with a red bomb. The bomb exploded and the creature evaporated leaving only a red cloud behind.

"Perfect job!" congratulated Bruce.

Yellow Car parked in the mining town. Bruce inspected a house on the edge of the settlement. It was a simple wooden cabin. There were a lot of heavy, dirty tools.

"We're going to find gold," said Bruce.

They connected the Golden Rule to a near mountain. Hundreds of trees were chopped down and a big hole was dug out in the middle. Gold was quickly collected, melted and placed in the trunk.

"We've gained about sixty bars," said Adelaide.

"What are we going to do with it?" asked Bruce.

"Some will go to pay our bills," replied Adelaide. "The rest of it we'll save in a bank and we'll use it later"

They liked to think of what they could buy. Adelaide wanted more books. Bruce wanted more weapons.

"I need some money too," insisted Yellow Car.

Meantime Joe and Frank discussed a new rule, which was named the Green Rule. Frank's clone reported that green is a very popular color and students were very fond of learning about it in school. The problem was not simple to solve and resources were scarce. The two players waited for inspiration to strike.

Then Joe built the elaborate prototype and connected it to Welcome Bunker. The entire hotel changed into an oasis with a lake and palm trees in the center. It was simple to reach coconuts, bananas and oranges from the branches just above. Frank had a feast without even moving from where he stood and Joe followed along quickly.

Bruce, Adelaide and their clones had arrived at Welcome Bunker the same time. They were pleasantly surprised to discover the changes Joe and Frank made while they were away. Their eyes were wide open and their jaws dropped.

"Joe was testing the Green Rule," explained Frank.

"Excellent test," agreed Frank's clone. "The best we have seen until now. Other tests we have reviewed in school only build golf courses."

"Clones go home!" commanded Adelaide.

That day the players were delighted and relaxed among branches; talking little, munching much and hanging upside down. Having eaten enough and rested they dived into the fine lake, and swam, splashing and laughing.

It was a very warm and noisy day in Welcome Bunker. All rooms were full. The parties continued on until dusk. The tenants had fun, went to bed late, felt asleep quickly and had colorful dreams after.

Wednesday afternoon Yellow Car popped out off to an Olympic landscape where several teams prepared themselves for Blazing Night's soccer tournament. Twenty-four teams were selected to enter. Prestigious medals were awarded to those who scored the greatest number of goals in three one-hour games played one after another with ten minute breaks between. Twelve games continued simultaneously and winners faced losers later until each team completed three games. A hundred thousands spectators in twelve soccer stadiums and millions watching remote monitors waited anxiously.

It was not difficult to figure out the team that Adelaide, Bruce, Frank and Joe supported: Purple Rats, an assembly of ambitious clones and dragons who dwelt in the desert. Bruce fantasized about joining this famous team, yet his physical education had ended somewhere in the minor leagues without any prospects of joining an intermediate, not to mention a major league. Florien Beige played left forward.

"The games are opened," Blazing Night announced.

These fantastic competitions started with shouting and yelling. Sometimes a roar from one stadium could be heard in the other eleven and motivated spectators to even louder applause.

Florien Beige ran half a field and fell on his nose. Purple Rats played against Rotten Singers. Purple Rats wanted to win all the games but they planned the highest score for the last game, the third one. Rotten Singers passed very well and advanced quickly.

"Goal!" shouted Rotten Singers.

"Goal!" shouted their fans.

Devastated by the loss, Purple Rats kicked the ball more accurately. After a few minutes Florien Beige ran to left and scored.

"Goal!" howled Purple Rats and their fans.

Who could play like Florien? No doubt he played as a top major league forward. Within the next fifteen minutes Florien scored twice.

Rotten Singers continued to press but failed to get through the Rats' defense. A mixture of dragons and clones in one team worked effectively and often helped to capture more awards. Rotten Singers sang beautifully and loudly but as the team consisted only of clones they showed no frightening tooth in their attacks.

Purple Rats advanced to a second game as three-to-one winners and played against Flying Robots. Flying Robots were swift and precise but as robots lacked the emotional factor that could be easily noticed whenever furious forwards of an opposite team kept scoring by executing surprising and spontaneous attacks.

"Goal!" roared Purple Rats and their fans responded.

This match ended with a seventeen-to-zero win for Rats. Adelaide, Bruce, Frank and Joe moved to another stadium where the last round took place. Spectators had to walk fast in order not to miss any part of the next game. An entire network of underground corridors with fast lanes of stairs operated without interruption. Hundreds of thousand spectators changed stadiums within few minutes between games. Only players were air lifted and transported to their destinations in less than one minute.

The last game was presumably against a weaker opponent but it was then that the Purple Rats claimed most of the goals. Florien ran and scored, ran and scored, ran and scored until exhausted he fell on his back immobile. The doctors sent him to the hospital. The final result of this game against the Flying Snakes stood at twenty-two to five, another win for Purple Rats. The Snakes shot splendidly from a far distance by hitting a ball with their long tails with such force and speed that it was sufficient to score if not defended before.

"Here are the final results," said Blazing Night. "The bronze medal goes to Stinky Frogs."

For a moment no one could hear anything but vigorous applause and fireworks exploding.

"The silver medal goes to Jumping Rabbits," said Blazing Night.

All kinds of neat silver stars exploded to a delight of the fans. Everyone congratulated the winners.

"The gold medal goes to Purple Rats," said Blazing Night.

Like an atomic blast a huge golden ball burst and sprayed everybody with golden sand so everybody, players and spectators, were covered with a cloth of gold, which stuck to their garments adding the flavor of a win to all at the end of tournament.

"Good night," said Blazing Night.

Thursday before departure to school, the clones had to listen to what the originals decided to order. The sun was just rising. All was silent. A chilly wind waved Frank's hair.

"You clones must be smarter," said Frank. "Are you not learning in school? Work harder! Go, and remember, don't be late!"

A few additional, clarifying morning instructions did no harm, Frank thought, the dummies would improve one day. But that was not certain.

Then the troops mounted Yellow Car.

"Ready to dive?" asked Bruce his friends smiling. "Hold on to your seats!"

"Let's go!" yelled Joe.

Yellow Car sank in Blazing Night's dark fluid. Soon they were catapulted out and landed on an orange planet. They found themselves racing with a colorful band of cars. The track continued from south to north with thirty turns that were specially bent to let vehicles speed up faster and to improve traction, perhaps making the race safer. Some drivers were clones, others were players. One could not tell the difference when the average velocity was thirty seven million knots and in which visibility limitations forced players to rely on detectors that in such speeds made numerous mistakes. Frank was driving. Adelaide was piloting. Bruce and Joe watched controls and monitors, seeking any condition requiring their attention. After recent upgrades, Yellow Car had no trouble taking part in such rapid races as its devices were modern.

"Speed up to forty," said Adelaide, "We've got twenty two cars in front of us."

"Done," said Frank.

"There's a red car on your left," continued Adelaide. "I want you to pass it on our next turn. Approach it from outside and then quickly cut to inside and speed up."

Frank slowed down to eighteen millions before turning and went to twenty-five when getting out off to a straight. He passed one car as planned and hit forty-one million knots a few seconds later. It was a thrilling experience. Players held their breath and waited in silence and listened to the engine roar.

"Sounds fine," said Bruce.

"Can we go faster?" asked Joe.

"I don't know," replied Frank. "Can Yellow Car take it?"

"Two cars to pass next," said Adelaide. "Go for it."

"I'll try," Frank answered.

"Trying is not good enough," Bruce shot back. "Make it happen!"

They drove at forty-seven million knots. Everyone stared at monitors expecting the unexpected. Yes, they passed two cars briskly.

Suddenly a strong east wind pushed Yellow Car to the edge of the racetrack.

"Turn against the wind," shouted Adelaide.

"I can't," said Frank.

"Joe, release a rocket!" shouted Bruce.

"I can't," said Joe.

Adelaide very leisurely crawled to the console and pressed the shiny, red button. Black smoke came out of Yellow Car and covered a valley nearby. Within seconds all players were wrapped in isolation layer and were kicked high up into space, well above the smoke. Parachutes opened and helped the players to land far from the track on grass among curious little Orange Dragons.

"Everyone's safe?" asked Bruce.

"Oh, yes," said Adelaide. "We're fine, but where is Yellow Car?"

"Isn't it dead?" asked Joe.

"I'm OK," whispered Yellow Car weakly.

"You look bad," said Frank. "Not like a car…"

In fact Yellow Car looked like a piece of junk. It was dragged along by Florien Beige, its frame held its body parts together. Yet it appeared that Yellow Car would fall apart at any moment. Florien dealt with Yellow Car gently as he got it to the shop near the beach.

The name of the car shop was Cheesy Fix. Yellow Car was parked there to be inspected by professionals. The damage was terrific. Yellow Car forced itself to think about its owners, their personalities, their fine looks and their concern about the vehicle. The water was calm. The grey ocean waves played a well known tune. Yellow could not stop imagining how nice it would be to again have them touch its controls, shoot enemies and run away fast, having its trunk loaded with treasures when suddenly a spark killed that thought.

"Blazing Night," said Bruce. "Get all of us to Welcome Bunker now!"

So she did. Her blaze was orange and gold. She transported them out of the game and left them in front of the hotel. Their dusty faces darkened their eyes, which gleamed like wet pebbles on a beach. Then she left, leaving her crimson shade and seaweed scent behind. They counted their bruises, washed themselves and went to bed.

Next morning the clones appeared to lecture their originals in the front of Welcome Bunker.

"You need to treat us better," said Frank's clone calmly.

"Why so?" asked Frank.

"The circuit part of your body may break under stress when we are dissatisfied with the flesh parts acting funny while the original makes an unfair comment," replied Frank's clone unhurriedly.

"Bruce, could you explain?" asked Frank.

"The Black Rule is primarily responsible for creating clones. The newer the version of the rule that is applied the more options and features clones get," explained Bruce. "After the creation process is complete, clones continue to learn and the original is the prime source of corrections. Regular laboratory upgrades fix only critical problems in order to avoid deletion of experiences," continued Bruce. "Any emotional experiences are most fragile and might be removed or damaged by an upgrade. Of course once a clone is declared dead, a

new clone could be created again using the newest version of the
Black Rule, but only those experiences which were saved could be
restored. Clone's experiences are held in various forms, including
diaries and any available logs," concluded Bruce.

"Why do I need to treat my clone fairly?" asked Frank moving his
fists in the air.

"Yours was created three months ago," said Bruce. "It does not
have much experience but was created, I think, using the newest
version of the Black Rule. This makes it more sensitive. It captures
your experiences faster and makes fewer mistakes."

"Clones please go to school now," Adelaide interjected.

"Ciao," said Joe.

Meantime the shop's engineers decided not to trash Yellow Car but
to rebuild its body and to install new monitors and modern sensors.
They promised to finish work within a few hours. Detailed analysis
of engines and of central systems revealed very interesting
information, which would be lost if the car were to be destroyed.
Piece by piece flew up in the air and made several circles high above
the repair shop then they merged and landed ready and fixed.
Technicians applied the Yellow Rule three times: firstly, to fix old
parts and to assemble it, secondly, to create new monitors and
sensors, and thirdly, to tune the fixed old unit with the brand-new
parts. They beautifully recreated its entire yellow body and renovated

its enhanced interior. After this hard work the project director ran million tests and verification procedures including phonetics and rhetorical quizzes. Technicians documented new acceleration, speed and temperature parameters and limitations. They wrote a recommendation guide following the best practices. Finally its weapon systems underwent sophisticated calibrations. This is how it became the finest artillery and rocketry available on the free market. All guns big and small shot more precisely and more quickly. Every rocket started faster and targeted better. If this was not enough for quality, entire sensor and monitoring systems functioned with substantially increased precision and when moving with the speed of two million knots could detect a fly from a distance of ten million miles. By twenty million knots, the car could detect a fly two million miles away.

After all that work was done, Yellow Car drove across Mediation Coast searching its owners. Finally, the players spotted it and waved. It slowly arrived and stopped in front of them as though it was smiling.

"How are you doing?" asked Yellow Car.

"We're fine," replied Adelaide.

"Let's go for a ride!" shouted Frank.

Joe drove deliberately around the town and back to Welcome Bunker. It was a pleasure to hear again the engines' power and to touch new controllers and monitors. Everybody laughed and joked.

They did not touch the new equipment yet but gently observed how it worked. Then Joe parked beside Welcome Bunker.

"Good night Yellow Car," they all said and went to their rooms.

Each player felt happier and more assured. Something warmed their hearts and made them smile before falling asleep: their car was parked nearby. A golden and yellow flame was also sneaking along beside the lobby but Yellow Car did not detect it. It was Blazing Night checking.

They woke up Saturday morning, walked downstairs, stopped at the door of the hotel lobby, and glared out at the sky. The murky rain would not stop. Black and brown droppings covered the streets and buildings of Mediation Coast, which was the small town where the players lived when gaming.

"What are we going to do about this damned rain?" asked Joe.

"We'll fly up to check on clouds," answered Bruce. "We may have to perform a cleanup procedure."

"Hurry!" said Adelaide. "I can not stand this wet dirt and what it is doing to Mediation Coast!" she mumbled.

They quickly boarded Yellow Car and took off. The clouds looked dirty and heavy. Players decided to evaporate water and to collect the rest. This seemed to be a clever plan.

"Joe, start heating," commanded Bruce.

"I'm running the Red Rule now," Joe reported.

Immediately the yellow and red fire jumped out of Yellow Car. Entire clouds changed color from black and brown to purple and blue. Water evaporated, and dirt got condensed to a big balloon which next fell to near mountains into a dumping field.

"We've done a fine job, thank you very much everyone," said Bruce.

They returned to Mediation Coast. The town looked different. Dark fluid patches covered several streets. Drivers carefully avoided it. Those who failed sank in the fluid and did not come back.

"Prepare an emergency cleanup and a rescue operation," shouted Adelaide.

"Joe, run the Brown Rule," commanded Bruce. "Adelaide, discuss a rescue procedure with Frank. We'll dive soon."

They became extremely busy and acted fast. Yellow Car collected all the patches, sealed them and destroyed all but one of them. They dived in and popped out in an arctic landscape, pierced by freezing wind. There White Monsters surrounded the Mediation Coast people, blocked their way and readied them for execution. The game instruction stated: "White Monsters are harmless." They did not appear to be such when they summarily executed innocent people and clones.

"Shoot them all!" commanded Bruce.

Adelaide did not have to be rushed. She fired several times and smashed five White Monsters in three seconds. Joe shot one. Frank drove at twenty million knots. Bruce opened a big gun and furiously discharged sixty rockets and melted the monsters instantly.

How the people and clones were saved and evacuated back to the town is history, well documented in the library archives of Mediation Coast. Adelaide, Bruce, Joe and Frank received a bronze medal for bravery in action. As a reward the Mayor of Mediation Coast presented the players with the key to the town which could open any door. Welcome Bunker's keeper decided to absorb any outstanding lodging costs and added two months of free accommodation for all four including free parking for Yellow Car. The game instructions were updated and any penalty related to it was revoked. Frank and Joe were heroes, for these two somehow knew better and sooner that the White Monsters and Dragons had to be shot freely. They could again enter Blazing Night every day they wanted. The town of Mediation Coast funded two bikes for Joe and Frank. They replaced the old ones immediately.

These events did not surprise Florien Beige. He knew how dangerous and frightening Blazing Night was. She acted swiftly and mercilessly. Conventional defense methods worked poorly and ineffectively. Florien had learned a lot about her. She imprisoned him once and kept him in a dark dungeon for four months until he luckily escaped through a small hole off into a desert. Blazing Night detested deserts. The heat and lack of fluids made her sick and weak. Perhaps a winning strategy would be to lure her to a dry and hot spot and to attack there.

Sticky Waffles was a restaurant where the four players enjoyed several meals and where the mayor and a bunch of other people organized a small thank you party on Sunday afternoon. Adelaide talked about it all evening and forgot to instruct her clone after school. It confused its tasks and decided to go to school instead of going home. Bruce had to catch it and directed it back. He also made sure the other three clones knew what to do.

Frank and Joe enjoyed their new bikes. They raced each other long hours around town and in nearby mountains. Joe was faster. Frank was more persistent. Uphill and downhill they rode their bikes.

Then they ended the day in Welcome Bunker. There players slept and dreamed deeply.

Twelve hours later Adelaide stood on the main porch of the Church. It was the most beautiful Cathedral she had ever seen. Two towers reached up toward the skyscrapers and stood like honor guards of the main door located between them. The faithful flocked inside that Sunday. Adelaide cautiously entered somewhere in its midst and sat beside a young couple quietly chatting.

It was the First Sunday of Advent. The Archbishop presided; the rector greeted the people.

The congregation sang a hymn and listened patiently to the readings. Frank could not focus on what was read and said so; instead he dreamed he was on a boat lost among furious waves. Ocean water kept covering his vessel until it sank and became a submarine.

The congregation sang,

"Look down from heaven and see,

From thy holy and glorious habitation…"

"Bruce, where are you?" Frank asked loudly. "Oops," he said softly after he realized some people looked at him from near pews somewhat disturbed.

The congregation sang,

"Yet, O Lord, thou art our Father;

we are the clay, and thou art our potter;

we are all the work of thy hand."

- ISAIAH 64:8

Meantime, Bruce studied large amounts of data that highlighted several cases of a little known virus. He had a persistent thought that Blazing Night had something to do with this. How often a virus was passed from one game to another and so entered a clone and infected the people next to it. These raised doubts about the effectiveness of security barriers surrounding the players. For the last few years the government had cut spending on valuable research, and as a result minute mistakes may have penetrated the Land of Games or even impacted the daily life of users. Bruce worked hard in order to better prepare himself for future challenges. He believed in unrelenting work, but often lost a sense of purpose, yet he kept studying, perhaps hoping to come back to a lost train of thoughts later. Today he learned that viruses acted viciously with no respect for persons either secular or religious. Doctors, biologists and chemists spent countless hours searching them out. Eventually they solved a problem by inventing the Purple Rule which when connected produced antibodies capable of an instant cure.

At the same hour Joe had fun playing with his brother in their apartment. They raced cars for several hours from one room to another, around a table, between chairs, and off to the bathroom. A newly met friend of theirs came by and the three of them started running and shooting each other until Joe's father asked them to stop it. He told them to behave and not to make noise, which regularly disturbed their neighbors.

Chapter 6: Boomy Dogs

A bitter, wild wind woke up the players and hurried them to breakfast at Sticky Waffles where the mayor and people of Mediation Coast waited. It was a thank you party, which had been planned several times before, but until now the four had not been available.

"How many times can one prepare a party for guests who aren't coming?" asked Adelaide wistfully.

"Until they come," answered the mayor laughing.

"Well, we couldn't make it before," Bruce explained apologetically.

A local band struck up a tango to the delight of all present. Waiters brought refreshing drinks, fancy food and colorful flowers. The most enticing was wiggling fish, which Bruce munched immediately. Everyone was talking and eating. When the clones arrived, Bruce instructed them quickly and sent them to school. Adelaide adored parties the most. She could party everyday; Joe and Frank would rather race or fight.

"How come you are such brave fighters?" asked the mayor. "You are so young."

"It's a matter of practice," Bruce replied modestly.

"We're eager to play," enthused Adelaide smiling.

"How much did your car cost?" asked the mayor.

"Three million pieces of gold," answered Frank.

"Who paid for it?" asked the mayor.

"We took out a loan," Bruce replied. "We often bring gold from the Land of Games. It pays our expenses."

"Are you never frightened?" quizzed the mayor.

"Oh, yes, sometimes," Frank admitted.

"I'm not," said Bruce. "Look, I'm like a terrible crocodile," he added.

"I'm like a swift scorpion," said Joe.

"I'm like a merciless anaconda," said Adelaide smiling.

"I'm like a bold vulture," whispered Frank in a menacing tone of voice.

They all laughed and talked about how pleasant and comfortable Mediation Coast was. It grew for many years yet not too much or not too little but just right. Inhabitants lived nicely. Tourism became a primary income. Players frequently sojourned here. Though it was a small town, its services were famous and well advertised in big cities. The mayor led his fellow citizens to continuous prosperity. They called him Andy Green. His widespread education included economics and mathematics. He built recreational and repair facilities for gaming equipment. Andy was a gambler. He wanted to take part in a gold expedition, in which players brought back treasures regularly and quickly.

"Could you take me with you on a gold trip?" asked Andy.

"OK," said Brucc. "Let's go now."

So they did. Yellow Car popped out in a wintry landscape. Hundreds of clones dug out minerals in freezing cold. The mining equipment drilled deeply into ice and rock. The hard working clones jumped into the pits, inspected the ground, and usually brought out a mineral sample, often a small golden nugget. Still, many monsters occupied vast tracts where bonanzas likely resided untouched. Clones always stayed far away from such places and only watched it from a distance.

"Regroup! Attack the beasts!" commanded Adelaide.

Four players shot the monsters swiftly and cleansed one bonanza entirely, leaving nothing behind.

"Joe, apply the Golden Rule!" shouted Bruce.

"Done," said Joe.

"Park here," said Adelaide.

Frank halted Yellow Car in the middle of a big hole. They got out and walked at a snail's pace. Suddenly a long-neck White Dragon punched the car from behind and threw sharp rocks at the players. Andy fell on his face and could not move and only his feet stuck out. Joe and Frank mounted their bikes and rode close to shoot. And they did. Within seconds nothing was left of the white creature. Adelaide heaved Andy swiftly and checked his fresh bruises.

"Are you OK?" she asked.

"I don't know," answered Andy.

Stacks of gold bars were piled in front of him, ready to be collected. Andy could not believe his eyes. He missed every detail of their mining operation when Joe connected the Golden Rule to the ground and how the minerals flew up, melted, and fell neatly placed one bar on top of another. He had only remembered that rocks hit him and pressed him flat to the ground.

"Where's Bruce?" asked Adelaide. "Joe, Frank, find him!"

They mounted their bikes and initiated the rescue. Joe shot anything that moved and did not look like Bruce. Frank inspected every corner of the big hole.

"Come back!" shouted Adelaide. "Andy saw Bruce talking to Yellow Car."

All five gathered around the Car and listened.

"Within five minutes a group of monsters will arrive here," warned Yellow Car. "You must escape. These are the Yellow Dragons and they never loose a battle."

Immediately everyone jumped into the vehicle. Yellow Car transferred gold to its trunk. Bruce decided to drive. He did so furiously and reached thirty million knots briskly. Close behind seven Yellow Monsters flew and watched.

"Faster! Faster!" shouted Joe.

Bruce pressed harder on the accelerator and clicked the black button in an attempt to make their car invisible. They became a cloud of yellow substance moving with a speed of thirty eight million knots. Yet the seven dragons reached it and caught its tail. At close to forty million knots no one could verify who was driving along. The Yellow Monster looked at its neighbors and tried to determine the enemies among them. Andy trembled and could not move. Joe and Frank gaped at monitors, buttons, and thought of something. Could they escape? Could Florien come to their rescue? Bruce held the speed steady. Either he was afraid to change it or did not know what to do. Everybody heard terrifying sounds of dragons' shrieking and skirling. Then Adelaide crawled to a left window and let out a confusion gas. It smelt terrible. The Yellow Dragons lost their balance and orientation. Their memories were wiped out and they could not recall defeat; they always won. Every Yellow Monster flew in a different direction. They did not know where they were coming or going. Totally lost, they traversed the sky like wind-tossed clouds.

Bruce slowed down to twenty millions and switched off the invisibility mode. All happily landed in Mediation Coast and counted gold. Andy Green took his share and went home silently.

On the day after these events took place, the clones and originals gaped at each other standing firmly in front of a wonderfully decorated Welcome Bunker. It was December. People were waiting for Christmas. The players mumbled to their clones about St. Albert's School, shaking their heads and shifting their hands. The term reports were expected to arrive soon.

"I'll see how well you're doing," said Frank to his clone. "I want the top marks!"

"Who doesn't?" interrupted Adelaide.

"Clones run to school now," commanded Bruce.

So they did.

"I plan to kill Brown King today!" said Adelaide glinting white light around.

The four players entered Brown Kingdom shortly after. Blazing Night tried to lure them to another destination but Adelaide held to their course and forced her to let them do what they decided instead.

It was Adelaide who learned how to impact Blazing Night's selections. Somehow they both bonded strongly and frequently fought little duels. Could Adelaide only win with Blazing Night? Could a merciless anaconda battle against a vicious, twinkling viper?

Yellow Car drilled a tunnel under the king's castle. Players readied themselves for a terrible battle. Bruce checked all the weapons. Frank and Joe held their guns firmly. Adelaide verified the position of Brown King and she followed his every move on the map monitor. Hundred monsters, king's bodyguard, constantly accompanied him wherever he went.

"We'll charge in a few minutes," said Bruce.

"Attack!" shouted Adelaide.

Yellow Car jumped out of the tunnel up into the air and reached fifteen million knots in several seconds. Players shot rigorously. Any creature within a range ran away leaving Brown King alone. He recognized the attackers and decided to negotiate a truce.

"Stop the car!" commanded Bruce. "We'll negotiate with the King."

They sat down and listened to Brown King utter his statements. He promised to free the clones in his kingdom. He also decreed that never again should any clone be forced to work for him. Every clone was free to move about within his kingdom. And finally, every clone was free to leave his kingdom.

Bruce smiled. He dreamed about victories when players brought freedom to those in need. He agreed to let Brown King stay on his throne.

Suddenly thousands of the Brown Monsters attacked Yellow Car from behind.

"Defend!" commanded Bruce.

"Run away!" shouted Frank.

Yellow Car could not accelerate fast enough to escape. A huge pile of Brown Dragons pressed it to the ground and left no room to either run or to shoot.

Florien Beige came swiftly to the rescue. He kicked the brown creatures and pushed them aside in order to destroy the pile sitting on the top of the car. However, new monsters arrived in great numbers and jumped into the battle, pressing ever harder. He could not clear them of the pile fast enough. The cracking sound of broken elements warned that Yellow Car was barely holding intact.

Brown King laughed and watched his enemies squeezed in front of him.

"Make a pie of them!" the king commanded.

Adelaide leisurely touched the yellow button of a remote detonator. Tons of red fire and poisonous gas appeared beside the brown throne, where she left a smart bomb. It cut of the king's tail immediately. It melted half of the piled monsters and badly confused everyone else including Florien Beige.

"Frank, drive fast now!" commanded Adelaide.

So he did. Yellow Car rounded the battle in circles at the speed of twenty million knots. The wounded, confused and almost dead monsters stared stupefied by their immanent destruction; one by one they were shot by Joe, quickly and precisely. Joe liked this type of operation when a talented soldier could smartly show of his battle skills. How he smashed the dragons and how he frightened them to death made history in the Brown Kingdom that day.

"Players, get out from the Brown Kingdom fast!" roared Blazing Night from the Dark and Warm Fluid.

They dived into the fluid, and drove quickly to Welcome Bunker where the clones waited already in Mediation Coast.

Ten hours of sleep helped to cool their tempers. Weak players dropped out of gaming quickly. Later they still enjoyed simulations, but did not enter a live game again. A new player received a license to own her clone and usually she purchased one on the free market. Players frequently acquired complex devices and fancy gadgets. Many created their own teams in order to cover up their own natural limitations and not to end their gaming careers prematurely. Once a team established itself and experienced several battles, it often registered its name in the Land of Games. Adelaide, Bruce, Joe, and Frank named themselves Boomy Dogs. Of course they all hoped that at least one of them would win and become a panhandler or perhaps they befriended someone who would become one.

Boomy Dogs entered the land of Blazing Night at 10 am local time. Within a few minutes they arrived at an island facing packs of Green Dragons. It seemed like a little surprise from Blazing Night on St Nicholas' day. Bruce could not stop rationalizing: why would they want to battle here?

"A treasure must be hidden somewhere in this green place," exclaimed Adelaide.

"I want it," said Joe. "I want the money!"

"Let's shoot a few monsters first," suggested Adelaide.

"I want it too," said Frank.

"Get your weapons ready," said Bruce.

"Yes, Sir," they all answered.

Frank drove Yellow Car that day. He flew up few miles and circled the island with the speed of twenty million knots. They monitored every creature down below.

"Attack!" shouted Adelaide. "Fight now!"

Frank lunged at the crowd of seven Green Monsters. Joe measured the distance precisely, and swiftly fired seven harpoons at the creatures below. Then Yellow Car glided gently into the middle of the tiny meadow. No live creature appeared at the front of them. The seven dead dragons lay side by side; harpoons pinned them to the ground permanently.

Bruce and Adelaide walked away and searched for meaningful clues. Frank and Joe stayed in the car and thought about hidden fortune.

"Why not run the Golden Rule here?" Joe asked.

"I'll connect it to a tree now," Frank replied.

So he did. A beautiful rainbow arched over of the rain forest and brightened the entire island. The light was strong and they had to close their eyes. A treasure chest appeared in front of the two.

Adelaide and Bruce wondered beside a little lake between two huge trees and watched for anything moving and dangerously alive. It seemed to be a calm place. Nothing was challenging them.

"We must be careful," cautioned Bruce quietly.

"Oh, yes, we must," agreed Adelaide.

She felt her way into a deep hole and vanished. Bruce did not see it and stared at the lake: something was tempting him to dive. He could not move his head aside.

"We'll swim now," he commanded and jumped into the warm water.

Both looked eagerly for cache. Bruce swam in the lake, Adelaide inspected the hole. Any corner could hide precious elements. But they found none.

Within an hour a hundred Green Monsters arrived nearby and captured them. Bruce and Adelaide became their prisoners. The

Green Dragons tightly surrounded the lake and the hole. No one could escape, it seemed.

Adelaide unhurriedly took her favorite gun and poisoned enemies instantly. It took a few seconds. The battle was over.

"The Green Dragons are naive," she said to Bruce. "I can kill them easily."

"Good work, Adelaide," said Bruce. "I haven't found any treasure in the lake."

"Neither have I in the cave," said Adelaide.

They walked toward Yellow Car. They understood that finding the game prize might not be as simple as shooting the Green Monsters.

Meantime, Joe and Frank collected their treasure and stored it in the trunk. This was what they won: three hundred fifty green diamonds, nine hundred red rubies, and one thousand small gold and silver weapons.

Bruce and Adelaide arrived at Yellow Car and interrogated the boys.

"How did you find it?" asked Bruce.

"We applied the Golden Rule to the forest," answered Frank respectfully as though he and Joe had done something wrong.

"Where was it?" asked Bruce.

"It got stuck in a bush," said Joe.

"But where?" said Bruce.

"Exactly where the end of the rainbow is," said Frank.

They laughed. This mission brought an excellent prize, a monetary reward. Yellow Car exited Blazing Night and parked in Portal Cove. Players rested under that huge root among rotten woods and polished rocks on the Portal Cove beach. Ocean smell and dull sun made this break pleasant and refreshing.

Their clones reported little progress at school. Bruce sent them home and told them to study harder next time. Heads down they heaved their flesh and metal bodies to where their originals lived.

Boomy Dogs carried their great win to Welcome Bunker and successfully transacted several business deals later that night.

Thursday morning the four ate breakfast. When they finished they furnished their equipment. It seemed that there was something in the air saying: fun, fun, fun.

"We must see this match!" shouted Frank to Blazing Night. "Don't try to lead us into any other destination only the soccer stadium."

"OK," agreed Blazing Night laughingly. "I'll arrange some of the best seats for you," she added.

Soon her shiny blaze faded away. Yet she left detailed directions explaining how to get to the stadium on time.

Jumping Rabbits played against Stinky Frogs and the winner took on Purple Rats immediately after. Blazing Night had constructed a new soccer stadium solely for this match and made it capable of holding two million visitors and fans.

Boomy Dogs got their tickets and landed on one of the moving segments of the arena. Their slice hung above the field and followed the action that often turned to focus on and stay close to the ball. They could also see the faces of the players and the bloodied parts of their limbs.

"Please welcome Stinky Frogs," said Blazing Night.

Applause came from the green side of the stadium and jeers from the long-ear sections.

"Please welcome Jumping Rabbits," announced Blazing Night.

Applause came from the long-ear sections and jeers from the green part.

They played fast. Frogs stuck to the ball and moved it ahead quickly. Rabbits lost their temper and one jumped on a Frog's head and smashed it against the grass. The whistle stopped the game. Two Rabbits were sent to the penalty box for one hour. Doctors took the injured Frog to a nearby hospital. Again Stinky Frogs attacked furiously and jumped over to a clear shot position. Yes, it was that famous dragon, Green John, who executed Rabbits instantly. He was the most eminent of the entire race of Green Dragons.

"Goal!" shouted the green fans.

Only the Green Dragons belonged to Stinky Frogs Incorporated, which was an excellent team that worked hard every day to improve their speed and play. They even used their tails to kick the ball harder. Everyone knew Green John and loved him for his roll in tail and kick scoring technique, which he had worked out two years ago.

That goal hit Jumping Rabbits in their weak spot. Could they come back? It was unlikely, for once Frogs had the advantage, they held to it.

Suddenly Franz Gray passed a few Frogs and kicked toward a left upper corner of the goal, and the goalie could not reach the ball, and everyone watched the ball falling...

"Goal!" shouted Jumping Rabbits' fans.

Franz Gray was the only dragon playing for Jumping Rabbits, others were clones and robots. He was a very experienced player, but even so Jumping Rabbits happily paid the costs. Many teams had wished to buy him. His last price was three hundred twenty five million dollars. But he had immense influence. He belonged to the great family of the Gray Dragons who lived in wastelands in which there are wrecks of mechanical and biological parts left unattended.

The game drew close to the final few minutes when five Frogs stepped on Franz and injured him. The main referee sent the Frogs to the penalty box for the rest of this game. The doctors took Franz to the hospital.

Now Green John ran across the field and a Frog passed him the ball. There was no offside, and he tail danced.

"Goal!" roared the green sections.

That was it. The game ended with a two to one win for Stinky Frogs. Jumping Rabbits cleared the field, heads down.

"Please welcome Purple Rats," shouted Blazing Night.

Beige sections applauded loudly and the green ones stayed surprisingly quiet. The match began twenty minutes after Frogs won and most of the Rabbit's fans went home.

"Florien, Florien!" Boomy Dogs shouted.

Immediately Florien Beige ran and scored.

"Goal!" roared beige sections.

"Way to go," shouted Frank.

Within ten minutes six Frogs jumped on Florien, squeezing him under. He knocked them down and damaged them severely. The doctors sent them to join Franz in the hospital. The main referee sent Florien to the penalty box for fifteen minutes.

Now Green John danced and kicked, but did not score. The Rats had a fine goalie, Big Ann, a lady dragon who fearlessly stretched herself to the limit anytime she blocked the ball. She certainly was the most expensive goalkeeper in the entire Major Soccer League in the Land of Games. Not many knew, but she was a favorite player of Blazing Night. Her biography highlighted a strong link to the uncommon group of the Silver Dragons. She may have worked in mines in the past.

That day the Rats worked well together. Either on defense or offense they passed and ran magnificently. They scored once more and so the match ended.

Boomy Dogs had forgotten about clones entirely. Very late they drove back to Welcome Bunker and even there kept talking about Purple Rats, how the match went and how many players were injured. Singing, shouting and laughing, they celebrated until nearly dawn.

Friday morning crawled to Welcome Bunker. The excitement of the tournament lost its bliss. The players anxiously waited in the lobby. It was a meeting time.

Bruce always said that hard work and long hours was what it took to plan winning strategy in battles. Boomy Dogs learned this very quickly. These four players carefully studied every report brought by clones from St. Albert's School. Of course Bruce was the most diligent in his studies and planning. Yet Frank's clone explained the matter in the most accurate way he knew. All rules and axioms seemed to deal with complex and difficult to comprehend realities. Often they lost hope of winning against Blazing Night. Who could beat her, they asked? She was a triumphant viper who controlled uncharted parts of the Land of Games. No one could precisely estimate her strengths and powers. Her only weak part, they figured, was a fear of the desert, perhaps a fear of drying in the scorching sun, perhaps a lack of water to drink. Florien Beige acquired this knowledge before he became a slave of Bruce, months before a Beige Rule prototype was applied upon him in the desert. Could they trust Florien's experiences? Had not Blazing Night repaired her weakness already? Those were the questions to answer soon if they were ever to win a decisive, final battle for primacy in the Land of Games.

That day Adelaide dressed herself neatly in an elegantly tailored dress. Her short black hair stuck flatly to her head between her small, round ears. Her eyes were deep brown and frequently glinted white light. Bruce had a big head, big ears, and a big ruddy nose. When he talked his lips moved quickly and revealed his two sharp teeth. His eyes were gray. They looked chic as though they tried to fool Blazing Night into believing that they were nice and gentle. Joe and Frank also wore clean, slick garments. Frank had azure eyes and ash blond hair. His nose was bent sideways. Joe was the smallest of all of them. His hair was fox red, and his eyes were bronze. He could bite very swiftly and hard.

Officer Brenda Hawkins, a tall handsome blonde, agreed to join their discussion and to help them in questioning Florien Beige, Yellow Car and Blazing Night.

"Who is your master?" she asked Florien.

"Bruce is my only master," replied Florien Beige.

"Who is your master?" Brenda Hawkins asked Yellow Car.

"My owners, Boomy Dogs," replied Yellow Car.

"Who is your master?" Officer Brenda Hawkins asked Blazing Night.

"I am the Master of the Land of Games," replied Blazing Night.

Brenda did not ask any more questions and left to pursue her regular duties. Blazing Night left them too. Meantime their clones had arrived after school.

"Florien, do you like Blazing Night?" asked Frank's clone.

"I hate her," answered Florien.

"Why?" countered Frank's clone.

"She's imprisoned me before. She wants me dead now. I know, she's a cold blooded murderer," he said defiantly.

"Yellow Car, do you like Blazing Night?" asked Frank's clone.

"No, I don't like her," answered Yellow Car.

"Why?" asked Frank's clone.

"She created me from a bicycle and sold me to Frank for a profit," said Yellow Car. "She ran the Yellow Rule and connected it to her Dark and Warm Fluid. This is how she created and sold me."

Bruce sent the clones home. They had learned a few important details that day. This inquiry was supposed to help them to understand Blazing Night better and to win faster. Every new bit of information led Boomy Dogs to the conclusion that they had to study and experience more before being even close to an equal battle against Blazing Night.

"Here's my version of the tale of Blazing Night," Bruce continued. "She was built about one million days ago in a small garage lab by a sick genius inventor. She incrementally learned to survive and settled in a network as an antivirus software that everyone installed on their systems. So disguised she grew to be independent. She had learned to live in a distributed environment, practically reviving herself and growing constantly. Any time she was wiped out, she revived herself once systems connected back to networks. She avoided any government or military organizations, but chose to dwell in gaming consoles instead. She even created several popular games. Innocent players unknowingly multiplied her everywhere. Their intelligence fed her for many days and that made her what she is today. Blazing Night decided not only to conquer the Land of Games but also the rest of the living space," concluded Bruce.

"We'll kill her," declared Adelaide.

"So we will," Joe and Frank agreed together.

That night was calm.

Saturday morning the originals instructed their clones in a precise, basic way in the lobby of Welcome Bunker.

"Take it easy, clones," cautioned Bruce. "We are close to Christmas, and our winter vacation begins within ten days."

"Yes, we will," responded Frank's clone. "We will prepare our reports for you soon."

"We'd like to learn more about the Beige Rule and the Green Rule," Frank advised. "Keep your notes up to date."

"We'll do that," assured Frank's clone.

"Now you may go to school," said Bruce.

So they did. Boomy Dogs boarded Yellow Car and dived into the Dark and Warm Fluid. They let Blazing Night select the destination of their voyage that day.

They waited and waited but did not pop out onto the shore. Rather they moved constantly in the fluid. Yellow Car warned of six monsters coming from the south.

"Speed up, Frank," commanded Bruce.

Yellow Car hit eight million knots. Visibility was poor and they relied fully on the car's sensors and its motion detectors.

"They're behind us," reported Adelaide. "Let's fight."

"Frank, you turn around and Joe, you shoot them quickly!" Bruce ordered, his curly hair shaking nervously.

The monsters looked like blue whales. They opened their jaws to bite Yellow Car, and they almost did. Frank drove faster but could not reach ten million. He directed the car inside one whale. Bruce discharged harpoons. The whale froze with its jaw open for a few seconds, long enough for Yellow Car to get away and Adelaide to leave a small mine inside. The monster was blown to bits. Its fat blue parts mixed with the fluid and blocked other whales for a minute.

"Regroup and charge again," ordered Bruce.

Boomy Dogs readied their weapons and targeted the five remaining whales.

"Fire!" shouted Adelaide.

Batteries of laser beams struck the monsters and flipped them upside down. Adelaide gradually attached small mines with large airbags and then pressed the trigger. Frank drove as far and as fast from the battle as Yellow Car could go in this fluid. The airbags ballooned and dragged the animals up to the surface. The mines exploded and broke each monster to bits. Finally, balloons popped out of the fluid but nothing was attached to them any more.

"Great job," congratulated Bruce.

"It's a standard operation," replied Adelaide. Her small lips smiled mildly momentarily displaying her perfectly white teeth.

"Thank you so much," Bruce responded and flashed his alligator smile in return.

Joe and Frank watched this underwater world with curiosity. They kept staring at monitors and spotted the little Purple Dragons. These swam stealthily among seaweeds and when a small fish appeared they instantly swallowed it. All of a sudden millions of Purple Dragons attached themselves to Yellow Car. Their jaws opened and they bit it firmly.

"We're in a big trouble," exclaimed Bruce.

"Yes, we are," agreed Frank.

Joe looked into the car manual and found an interesting procedure.

"I'll nail them now," said Joe, scratching his red hair.

This was a brand new defense feature that was added during the recent repairs. Joe stepped on the switch pedal.

"Nail them!" Joe shouted.

Yellow Car's surface changed color to orange and pierced every dragon with three nails, which spit a paralyzing poison into their tiny purple bodies. Frank drove faster and accelerated the car to fourteen million knots, a safer speed. Frank and Joe giggled nervously. This was a delicate and successful maneuver.

Boomy Dogs directed Yellow Car back to Mediation Coast where their clones waited.

"Good afternoon," said Frank's clone.

"Clones go home!" said Bruce.

The four players drove to Portal Cove where they rested after their battles. Florien Beige waited patiently on the beach beside rotten woods, colorful pebbles and sharp rocks.

"You have to warn your government," advised Florien. "Blazing Night plans to invade Earth!" he continued.

"How do you know?" asked Bruce.

"Master, your servant dug out a pit in his desert home and found water where it had never been found before." Florien Beige bowed down to Bruce and continued, "Your servant suggests that such a situation may imperil our hot, dry advantage, and thus weaken every corner of Earth's defenses against Blazing Night."

"Can we evaporate the wells?" asked Adelaide.

"I don't know," replied Florien.

Later that night Boomy Dogs drove to Welcome Bunker and stayed there.

Bruce warned the authorities. He sent a message:

Mediation Coast, Saturday, day 2,007,851

To Officer Brenda Hawkins,

Dig out wells in deserts and search for any trace of Blazing Night. She is ready to conquer Blue Earth.

Sincerely,
Bruce Maxwell,
Boomy Dogs.

At eleven in the morning of the next day High Mass began in the Cathedral. Sunday instructions were plain and helpful. Adelaide learned how to sing a hymn from a book. It was not easy. Some hymnals were older than others; words and music did not always match. Often she mixed text when looking at rows of words just below the notes. She tried hard to project a finest tone with the precision of a professional artist.

That day they sang,
    "No evil will befall the man who fears the Lord,
    but in trial he will deliver him again and again."
            - SIRACH 33:1

It moved her a bit and she smiled.

Adelaide's father went to confession. He did so once a year but never prepared himself very well. Often in the lineup he thought of sins he had committed and what to say to the priest. However, it was a spontaneous expression of contrition rather than a detailed study of one's faults and failings.

Frank listened to the lessons and the scripture. He even asked his father a few questions. He wanted to know when they could leave the church and what were they doing after. His azure eyes scanned every nail in the pew.

Frank's mother strolled along the mall and bought ladies' stuff nobody but her cared about. She loved shopping. Her wardrobe was extensive. Everybody watched her legs.

Meantime Bruce studied cases of virus attacks dated from a million days ago to the present. There were forty seven thousand kinds of viruses. The subject lacked detailed documentation. Some articles missed entire sections dealing with the methods of handling viruses and mentioned only tool names to abate it. Carefully designed security levels disallowed a free access to all related information. Only one report mentioned a case of an antivirus actually creating and populating a newer virus. It was the case of a Trojan horse attack. The antivirus, once installed, attacked a near system three years later. Bruce thought about a chain of events and how one made sure a demand for one's services stayed high. He studied a competition law and how some firms broke it and later paid relatively small penalties, often ending as the one and only provider of an important service. He knew that feeling of not knowing better, of being blindly content and of not having an opportunity to learn the difference.

Bruce's grandmother worked on her book. She wrote about the struggles and battles she endured playing games many years ago before Bruce's father was born. She loved her grandson very much.

That afternoon Joe and his brother Fred watched a cartoon movie about St Nicholas, who was lost in the snow, but was rescued soon enough to deliver Christmas gifts on time.

Fred prepared a board game he had created for the entire family to play. And they did.

Chapter 7: Desert Battles

"Joe, you're so tiny," said Adelaide. "How come you're such a brave warrior?"

"I practice a lot," replied Joe smiling. "Fred and I have fought with our toys since we were born."

"Is Fred a player?" asked Adelaide.

"I don't know," Joe answered.

Their clones were already in school. Boomy Dogs expected Brenda Hawkins to arrive soon. She did, with a team of players called Happy Campers. Then she introduced them to each other. These were members of Happy Campers: Jonathan, Alfred, Alice, Anne, Emma, Kevin, Rachel, Keenan, Kate, Fred, Arnold, Sigfrid, Valerie, Aaron, Jose, Alberto, Anna, Carlos, Olivia, Giovanni, Jorgen, Dimitry, Nadia, Ali, Aiko, Chen, Chow, and Li, twenty eight players.

"Happy Campers will work with Boomy Dogs and move to Welcome Bunker in Mediation Coast, effective immediately," commanded officer Brenda Hawkins. "After a short introduction both teams will investigate Blazing Night's whereabouts and verify every new well in the deserts for any sign of her there. The chief of this operation is Adelaide Smith," Brenda concluded and left to pursuit her other duties.

Many players surprised each other. Fred talked happily to his brother Joe. Bruce chatted with Chow, his classmate from St. Albert's School grade nine. Frank discovered that Giovanni was an altar server in their church. Others recognized several faces but were not sure where they had met them before. All of them remembered their vehicles: Yellow Car and Blue Bus. These two machines had battled the dragons and monsters under the command of Officer Brenda Hawkins as part of that fifty-ship mission weeks ago. That mission took place somewhere between Blue Earth and Mars.

"Board your vehicles!" shouted Adelaide. "We're leaving now," she commanded.

So they did, and drove to a desert. Yellow Car and Blue Bus were moving side by side.

"Speed up to twenty million knots and demolish every well you see," commanded Adelaide.

The entire assignment depended on precise bombardment of desert wells. Joe liked it a lot. He took little time to target and blow it up. One by one hundreds of wells exploded and their water evaporated. Happy Campers effectively did their part of this operation. It required excellent coordination of driving, targeting and shooting.

Suddenly twenty Brown Dragons jumped out of one well and shot at Blue Bus.

"Fly up in a double circle formation!" shouted Adelaide. "Attack!" she bellowed once the two vehicles were in the air.

Twenty monsters melted instantly. Another five hundred jumped from the well.

"Joe, destroy the well!" shouted Adelaide.

"Yes, Sir," Joe responded.

He pressed a blue button to discharge a small rocket. It circled the well twice, and dived deeply into it. It boomed loudly. Miles of red and blue gas and dust erupted like a volcano from the well. They could not see through it. The Brown Dragons regrouped and prepared to counterattack. They jumped on Yellow Car and kicked Blue Bus.

"Retreat!" shouted Adelaide.

It was the last chance to escape from the muddled and bloody fight. They ran away. Adelaide slowly triggered the explosion of a smart mine she had left behind by the well. They had not seen such a colorful boom before. There was a white and yellow core in its center, a pink cloud around it, and finally a third ring of bluish and heated fire like lightning. Every living creature had to evaporate in the heat and electrical discharges.

"Mission accomplished," said Adelaide. "We return to Mediation Coast now."

Bruce sent a report:

Mediation Coast, Monday, day 2,007,853

To Officer Brenda Hawkins,

The Brown Dragons are associates of Blazing Night and were found in one well today. Hunt down and shoot the Brown Monsters everywhere.

Sincerely,
Bruce Maxwell,
for Boomy Dogs and Happy Campers.

Players sent their clones home very hastily and had a small party in Welcome Bunker. Boomy Dogs and Happy Campers operated efficiently and effectively together. If they were not at war one could say they had fun competing.

Those combat events had led to a closure of the Land of Games. Only players on a battle mission entered it until further notice. Mediation Coast became a command center of secret operations directed to fight the Brown Monsters and Blazing Night. Sports tournaments had their schedules updated and several were canceled. Fans decided to protest against the government and demolished five shops and the parliament building. More chaos could be expected unless the games were back and on.

Boomy Dogs and Happy Campers performed regular military duties. There was no time to instruct clones. Teachers issued more than the usual homework alerts and warning notes to parents.

Chow's clone fainted in her class when Jordan Blake, her teacher, asked her to explain the Beige Rule and the Green Rule, and compare the two. They sent her to a regional hospital. A doctor recognized a clone and reported the incident to St. Albert's School. The principal of the school asked Chow's parents to come and explain. Her parents knew nothing about her owning a clone. Her parents did not understand how a young girl could take part in battles instead of going to school and studying.

Two clones, Aaron's and Alberto's, bullied other children. They stole sandwiches, candies, cookies, and pencils. They pushed and kicked smaller students and made fun of them in front of everyone else. Bystanders frequently laughed. The suffering students were dismayed. Clones could not be caught in their acts. They lied regularly to cover up any shortcomings. Teachers listened to their lies and understood them as truth.

Kevin's clone pushed his family into a crisis. His father left home and did not come back. His mother hysterically and repeatedly called police to find him. His father did not want to come back. He stayed in a pub.

Arnold's and Sigfrid's clones wore military uniforms and carried weapons. They were grounded for one week of house arrest. They stayed in their rooms and wore their uniforms.

Nadia's clone refused to attend her school. Nadia's father could not convince her otherwise. She stayed home, closed her room and watched one movie all day again and again.

Frank's clone figured out that he should be instructing other clones. His right hand was raised high. He talked to every clone. His head swung left and right. He asked them to stay calm and to follow procedures they still remembered.

"Don't be late," Frank's clone cautioned Giovanni's clone. "Get the top marks for this term," he challenged Dimitry's clone. "Hide behind others," he advised Ali's clone. "Do not talk to your teacher but only smile," he explained to Chen's clone. "You must be smarter," he commanded Jose's clone. "Write detailed reports," he said to Li's and Aiko's clones.

Within a few minutes Frank's clone sent others to their schools. All listened to him and nodded yes. There was calm.

"Clones go to school," he said to them loudly and went to St. Albert's alone.

Once clones arrived from schools he stopped them. His right hand was high. He looked at them straight.

"Clones go home!" he ordered.

So he and they did.

Meantime, every player was traversing the desert all day long killing the Brown Monsters. It was reported that about twelve thousand monsters and dragons were slaughtered that day. They stayed up late and fought the brown enemies.

Later in the night Boomy Dogs and Happy Campers tirelessly discussed their daily duties. How they worked together and shot so many Brownies. How Frank drove very fast in circles and Joe stabbed those dragons to kill them immediately. How Giovanni and Emma shot twenty monsters at once. How Keenan, Fred, and Chen smashed forty dragons with one smart bomb. How Alfred and Alice evaporated twenty two wells. All these stories were recorded in the history of Mediation Coast, and were carefully archived for future generations to learn and not to forget.

A chilly breeze blew over the heads of the lone duplicates in Mediation Coast Wednesday morning. Frank's clone became a point of contact for about seventy other clones. The originals took part in battles and had little time to instruct their own.

"Good morning clones," he greeted them before school. "Go and learn, my friends," he concluded and moved his nostrils up a bit.

Somehow that attitude made other clones act smarter in schools. They performed better, and teachers did not complain any more. It was not clear why, but no one was asking.

Meantime five teams, along with Boomy Dogs, fought against seven hundred Brown Dragons and the battle continued unresolved. Boomies and Happies had already killed one hundred monsters but the Brownies kept coming from wells on Savvy Plate, a desert far away from Mediation Coast.

Suddenly a viper-shaped vehicle popped out of one of the Savvy Plate's wells. It was full of Brownies holding weapons on one side of it and stuck to the other. The viper flew up to the air above all players. Now jets of fire came from the creatures above or below. Adelaide was in charge of all five teams.

"All crafts dive into the wells and shoot the viper above!" she shouted.

They were scattered among the Brown Dragons jumping on top of them and the fire coming from above. The end was dreadful, it seemed.

Florien Beige ran to rescue them. He cut the viper in two and melted the Brownies, which piled on the top of Adelaide's party of five machines. It was enough to permit a retreat to a safer southern corner of the Savvy Plate desert. There they ran.

"Fly above the viper and charge with rockets and mines!" commanded Adelaide.

She slowly started detonating mines left behind. Twenty paralyzing mines exploded. Next melting mines ignited and spattered half the Brownies on the ground. Vehicles turned up to the air and became ships and crafts way above the viper.

"Fire!" shouted Adelaide.

Hundreds of red, blue and yellow rockets and bombs exploded on the surface and above Savvy Plate. A brown melted mass slithered into the wells. The viper vehicle, cut in two, ran one side east and another west, regrouped, changed color to silver, and attacked with lightning, bringing down Adelaide's battalion, completely confused and disoriented. Florien Beige was kicking the western part of the viper and blasted all Brown Monsters stuck to it. The eastern part entangled Florien surrounding him completely. This momentary delay helped Adelaide to send a distress signal to Brenda Hawkins.

Florien Beige, the dragon, fought with the viper. The eastern half seemed to be stronger and did not let Florien breath. He bit it, kicked it, punched it, but could not get it of his body.

The whole army of thirty ships, grouped and directed by Brenda Hawkins, was due to arrive soon. Messy and bloody mixtures tidily paralyzed most of the fighters. Bruce could not stop imagining them perishing while the rescue mission, trying hard to reach them in that crowd, missed by just two minutes.

"Attack!" commanded Adelaide.

But Boomy Dogs and Happy Campers together with another three teams sunk into a brown mass and did not move. Slowly Adelaide detonated every mine she had set and was shooting at the eastern part of the viper. Frank, Bruce, Joe and everyone else did likewise. Somehow silver armor protected the viper and it stayed alive and squeezed Florien harder and harder.

Meantime, far distanced from that struggle, new orders were issued by Brenda.

"Within five minutes we'll join the Savvy Plate battle, prepare your weapons," Officer Brenda Hawkins commanded her thirty-ship crew.

Help was coming.

Alberto, Joe, and Frank mounted their bikes, separated themselves from the rest, and drove on the eastern viper. They poked it with their sharp nails, which fired within seconds making small holes. They attached, left behind, and detonated sixty portable mines. Then they returned back to their crafts, still grounded on Savvy Plate.

Florien Beige, the dragon, fought bravely. Mine explosions helped him to regain his breath and strength long enough to renew pummeling the eastern monster. The western part melted already and left behind a brown mass flowing toward the wells.

"Burn the brown mass," commanded Adelaide.

All capable vehicles fired at the brown mass. It changed color to black and slowly evaporated leaving nothing but sand behind.

"Seal the wells!" shouted Adelaide.

So they did. Seeing that, the eastern viper released Florien and dived to a still opened well instantly. Then it was gone.

The Battle of Savvy Plate was over. The troops attended to their bruises. Officer Brenda Hawkins began inspecting every well. Yellow Car, Blue Bus, and three other vehicles were parked in the finest repair shops, where they could be repaired and enhanced.

The exhausted players went back to Mediation Coast to Welcome Bunker. None of them had taken part in a battle like this before. They were shocked. Such lightning, they experienced, had never been seen before. Vehicles looked like burned candles. Players could not talk, but silently went to bed and slept.

Florien Beige vanished and no one knew to where. Bruce suspected that he had dived into the well behind the back of the eastern viper; the dragon did not let go his prey.

Officer Brenda Hawkins and her army of thirty ships continued to seal the wells until all were closed. They worked the entire night. When the sun came back, there was only scorching sand. Burning piles of carcasses, looking like islands, were turning to ash and soon vanished into the swirling wind.

Next morning the warriors met at Sticky Waffles and plotted strategy. The gloomy future had to be turned into a victory, most of them guessed. But they knew not how. Their dim eyes searched for a leader to come and tell them what to do.

"We must kill Brown King," suggested Adelaide.

"I agree," said Officer Brenda Hawkins. "We'll do that!"

The two chiefs of the extraordinary army of players discussed a plan for defending Blue Earth from the Brown Monsters and Blazing Night who, as some said, was like a vicious twinkle viper. They wondered who could battle her other than this splendid army of gamers.

"If we fail, who else can come and stand up to Blazing Night?" Brenda challenged.

"We must move the center of our battles out of deserts into her territory," said Adelaide. "The Brown Kingdom is a perfect place to strike."

"I agree," replied Officer Brenda Hawkins.

"Let's divide our force into two groups," continued Adelaide. "One attacks from underground and the other from above."

"I agree," said Officer Brenda Hawkins. "Yes, you're right."

"I propose twenty ships go with me and twenty with you, Brenda," said Adelaide. "I'll attack from underground. Let's do it now."

"I agree," said Officer Brenda Hawkins. "All crafts and ships prepare to depart immediately! We leave in three minutes."

The entire forty ships and crafts dived to the Dark and Warm Fluid with a speed of two million knots. No one asked Blazing Night for directions. Everyone drove straight to the Brown Kingdom. There they separated. One party dived underground and the other flew way up into the air.

"Attack!" hollered Officer Brenda Hawkins.

Twenty colorful ships started shooting at the castle where Brown King used to live. A tremendous fire and quake followed. The western and the northern walls fell down. The Brown Monsters, in thousands, shot up toward twenty crafts of this splendid army circling at about twelve million knots and shooting down.

"Send bombs and mines!" shouted Brenda.

The Brown Dragons divided themselves into two parties, one defended the king and the other ran to a nearby city. They came back on top of the eastern viper and fought a battle in the air high above the castle where Brenda and her ships circled. Lightning filled the sky and the splendid ships fell to the ground where the rest of the Brownies jumped on them and kicked them.

Then Florien Beige, hidden beside the viper, melted every Brown Dragon stuck to it. They all fell to the ground. The vile viper injured herself and could not fly any more but her armor held firm on her body. There was a huge brown pile, and in it monsters live and dead mixed with the splendid army of Brenda's ships.

It was the moment to counterattack. The merciless army of Adelaide's twenty ships and crafts popped out of the ground and flew high in circles, slowly detonating bombs and mines that quickly and precisely smashed the Brownies. One could almost see the aiming and precise shots by Joe, and Yellow Car, driven by Frank, circling above like a vulture. Blue Bus and Happy Campers drove slower but Giovanni and Chow blew artillery continuously and left the Brown Monsters lifeless behind. Other crafts and ships followed and bombed the Brown Kingdom with the precision and strength of merciless players.

Soon they made a big breakthrough, allowing Brenda's army and Florien Beige to escape and to join the main offensive soon after. Forty ships and crafts shooting from above destroyed all brown defenses. The king and the eastern viper lay under the brown mass of dead flesh below.

"Burn it!" snapped Adelaide and smiled nastily.

This was Bruce's part to enjoy. He connected the Beige Rule to their bombs and rockets and sent them down. It worked well. Every brown creature changed color to beige and turned to dry and burning hot sand. However, the viper and Brown King escaped.

The rest was the history of the Brown Kingdom and how Happy Campers established a new government. Bruce helped them to draft a constitution, which was stored in their library. Clones ruled themselves under The Right to Freedom Act for Clones and the Right to Make Decision for Robots. The Brown Kingdom became a land for tourists and players to visit, to play in and to enjoy it. Some said that what was left of the terrible Brown King and his Brown Monsters were cookies called brownies, which were made available on every corner of the city's downtown.

That day no one learned what happened to the viper and the king, for their bodies were never found and several clones reported some brown mass had leaked into the Dark and Warm Fluid.

"Mission accomplished," congratulated Brenda Hawkins. "All crafts and ships return to your territories now."

Happy Campers stayed in Welcome Bunker with Boomy Dogs that night. Together they enjoyed fine food and refreshing drinks at Sticky

Waffles. A little party it was as resources were scarce and the fear of an unknown future haunted them.

Friday, the military authorities ordered the continuation of the wells and desert inspections. Due to security concerns the Land of Games was not open to the public. Officer Brenda Hawkins reported all important facts of the Battle of Savvy Plate and clarified how the magnificent army freed the Brown Kingdom of their king and his allies. Boomy Dogs instructed their clones and sent them to St. Albert's School after. Players had noticed a change in the clones' behavior. They acted and talked more effectively.

"How are you, Blazing Night?" asked Adelaide.

"I'm fine, thank you," replied Blazing Night.

"Are you ready?" asked Adelaide.

"Oh yes, I am," answered Blazing Night quietly.

Yellow Car plunged to the Dark and Warm Fluid. Frank drove nervously. They popped out in a deep green forest. Blazing Night selected the destination that day. Did she want to frighten Boomy Dogs or Adelaide only? She, a ruler, remembered well their battles and did not disclose any information about her role in it. She, a vicious, twinkle viper, did not appear as confident as before. Her voice was softer and quieter than it had been days ago.

"Prepare your weapons," ordered Adelaide. "We'll shoot a few monsters soon."

"Yes, Sir," answered Joe.

"We'll do all right," Bruce added.

Frank drove through trees and bushes leaving behind a trail of smashed wood and crushed greenery. Next they flew up and saw that it was an island. Yes, this was the very island where they had won a treasure nine days ago. It seemed like weeks, not days, ago.

"We'll drive to a lake Bruce inspected before," said Adelaide.

The water was warm.

"Let's check the bottom of it," Bruce suggested.

Bruce could not swim at the depth of the lake bottom, so Yellow Car, now a submarine, scanned its entire floor using its fine sensors and motion detectors.

"Frank, turn left, drive toward the big cave," commanded Bruce.

They slowly moved inside and penetrated the interior of the cave. Small Green Monsters stuck to Yellow Car but let it go when Joe quickly treated them with the nails. Bruce opened his mouth and showed his teeth, smiling like an alligator waiting to snap its jaw onto a juicy meal. His intuition led him here and now he knew why. They had landed in a treasure cave. Their car lights glistened on hoards of gold, silver and gemstones and reflected off of the amber ceiling. Gold, silver and gemstones covered it entirely. All this waited to be collected by the brave and lucky. Bruce gazed candidly at a game. He looked with reverence at bars of gold, signets, tiaras, crowns, robes, vessels, jars and weapons. All was richly decorated with colorful gemstones and silver touches.

This was what they brought back to Mediation Coast from that trip: a hundred twenty bars of gold, three hundred signets, seven tiaras, fifty crowns, seventy robes, one thousand five hundred vessels, six hundred jars, and three thousand weapons. All that beautifully crafted and decorated with precious stones: diamonds, rubies, amethysts, and nephritides. The amount of wealth, the Boomy Dogs were worth, had substantially exceeded other players' accounts that day.

But the enemies were working hard on Mars. Blazing Night regrouped her forces and prepared a surprise, deadliest and hideous Fierce Snake. Fierce Snake organized entire armies of monsters and dragons. She readied an invasion on Blue Earth for next week. Her forces included two armies: three hundred Brownies, which escaped into the Dark and Warm Fluid, led by Brown King, and four hundred multi-color monsters, which escaped to Mars after a battle with Brenda Hawkins fifty days ago, led by Fierce Snake herself. She was a silver armored snake which previously was known as the eastern half snake which fought in the Savvy Plate battle. Blazing Night repaired her and rebuilt her might after Florien Beige cut her in half. This she did when Boomy Dogs busied themselves collecting green treasures one day ago. These two armies dug a tunnel in Blue Earth and planned to attack in three days. Blazing Night executed detailed equipment verifications and upgrades that helped all monsters to regain their powers and made them faster.

"We'll attack from our tunnel," ordered Fierce Snake. "Brown King, you start our offensive"

"Yes sir and what will you be doing?" asked Brown King.

"I'll lead the second army to ambush all players," explained Fierce Snake. "We'll attack later, once all players join our battle."

"Are we strong enough?" asked Brown King.

"That's to be seen," replied Fierce Snake.

"I equipped you with the newest weaponry known in the Land of Games," advised Blazing Night. "I also added secret venom to be used by Fierce Snake."

"We're about five hundred monsters and dragons," said Fierce Snake. "I'll try to bring more troops from other games but this will take a few days."

"I'll also arrange additional enforcements," said Blazing Night. "We must prevail."

"But we have lost several battles before," Brown King noted nervously.

"That's a part of our plan… I let players think they're winning," said Blazing Night. "The real offensive commences in three days."

"Why was I not told that?" asked Brown King.

"Because I'm the master and you are my servants!" shouted Blazing Night. "Go and exercise your troops now!"

So they did. Two armies of Blazing Night continued to enhance their equipment and tested combat simulations on Mars. They verified their speed, strategy, and their offensive and defensive valor.

Meantime Boomy Dogs and Happy Campers lay flat in Portal Cove among rotten woods and polished rocks listening to ocean waves and smelling the coastal breeze. They thought how nice the weather was and what they would do during the winter break coming in seven days.

Sunday morning Frank went to the sacristy and met the altar servers there. He volunteered to become one, but he needed training. The rector promised him that one of the servers, George, would instruct him. His mother and father were very proud of him and were surprised by his courage to step up in front of so many people. They decided later to go to a restaurant and to arrange a small feast for the family to celebrate.

That morning Adelaide smiled a lot and wanted to be loud.

She joined the congregation and sang along the hymn,
"He who loves his son will whip him often,
In order that he may rejoice at the way he turns out.

He who disciplines his son will profit by him,

and will boast of him among acquaintances.

He who teaches his son will make enemies envious,

and will glory in him in the presence of friends."

- SIRACH 30:1,2,3

Adelaide performed well, she thought, and enjoyed the service even more. A fine soprano like her mother, she felt that she belonged there. Adelaide could read music and sing in four part harmony or in unison. She seemed to be convinced that the printed notes on white paper, marked and placed at their pitch, were somehow related to her favorites: parties and dancing.

Bruce worked very hard on an obscure document about prophecy and how some people could experience events that were later realized. Scientists did not find any rational explanation for this but often called it a process of intuitive repetition. Bruce also read about second thoughts, often quite unlike the original. They came to some people like a flash, in the middle of the night. Finally, he read a combination of intuitive repetitions and that blazing nights could, as the result of revelation, find of great importance. He read that it was one of those big inventors' ways of approaching a new unknown. These were difficult concepts to understand, and Bruce fell asleep while reading. His mother turned off his room lights.

135

Joe and Frank raced their cars for a few hours and fought one battle for another few hours that afternoon. They had to play slowly. Their father had worked hard all week and did not tolerate noisy games on Sunday.

Chapter 8: The War on Mars

Monday began delightfully. Clones wanted to learn more and prepared preliminary reports. They had five days of schooling before the winter vacation. The originals glared at their smiling doubles.

"Explain to me your newest findings about the Beige Rule and the Green Rule now," Frank ordered his clone and raised his right fist.

"Well, we're research them, and I'm writing a short explanation of the findings in my report," said Frank's clone.

"Go ahead and focus on military applications only," ordered Frank.

"The Beige Rule helps to find a better strategy or weapon to use in a battle, why the Green Rule helps to construct a poison gas to deploy swiftly," explained Frank's clone. "Either rule is very risky...the results of their applications are unpredictable. Often one's personality impacts on a final form of it."

"That's enough! Go to school, clones," ordered Bruce.

Boomy Dogs boarded Yellow Car and dived into the Dark and Warm Fluid. Adelaide forced Blazing Night to land them on Mars. Adelaide did not know what destination to select but she somehow intuitively felt that Mars was a place to which Blazing Night did not want them to go.

"Frank, speed up," Adelaide commanded. "We're hunting armed bases."

"Yeah," Frank replied.

Then he drove Yellow Car, which was now a rocket, at the speed of thirty million knots. Frank chose an exploration path around Mars and flew a scanning pattern in small circles. This allowed them to detect and record any military progress. They studied meticulously every piece of incoming data. It took them one hour to collect an entire set of images of Mars. Yellow Car highlighted two locations where it indicated groups, which were moving in a way that was similar to military exercise. Yellow Car highlighted also two other locations where monsters and dragons assembled in the reservist like formation.

"We're loosing the war!" shouted Adelaide. "Bruce, alert Blue Earth now, quick!"

Bruce immediately sent the following emergency message:

EMERGENCY

Mars, Monday, day 2,007,860

To Officer Brenda Hawkins,

Two armies are ready to attack Blue Earth and another two armies are building up their forces on Mars today. Attack Mars now.

Sincerely,
Bruce Maxwell,

Boomy Dogs.

Boomy Dogs transferred all their financial assets and purchased combat robots and vehicles. Their orders clearly demanded that they be shipped to Mars immediately. Boomies set up a small but well armed base on the north side of Mars. There, in the dark, chilly corner of the planet, they expected the equipment to arrive within hours. Not only were light devices and gadgets ordered but also bunkers and military headquarters.

"We will probably stay on Mars for the duration of this war," Bruce thought.

Very quickly they received information from Blue Earth:

Mediation Coast, Monday, day 2,007,860

To Boomy Dogs,

Stay where you are. This is war! All students and staff of your school have been abducted. The kidnappers are negotiating the release of the hostages. The school is surrounded by the police and an anti-terrorist battalion.

We have discovered a tunnel where the invasion from Mars is planned to take place. I will join you on Mars soon.

Sincerely,

Officer Brenda Hawkins.

Boomy Dogs learned that the real offensive was on Blue Earth. They realized that everything before had been just a strategic maneuver to blind players and to make them think they were winning.

"This is the work of Blazing Night," said Bruce. "She's using the same strategy she did when Frank lost and fell of the racetrack weeks ago."

"I agree," replied Adelaide.

Boomy Dogs did not wait for Brenda but divided their forces into two groups and sent them to fight the two big armies of Blazing Night.

Soon after that multiple explosions leveled the enemy's camps. Robots fought effectively but could not hold out for more then one day. Their assignment was to keep enemies busy, so they failed to invade Blue Earth.

"We can not loose this battle!" said Adelaide.

"We must come up with a brilliant idea," said Bruce.

"I have no clue," Frank confessed.

"We can try to prototype the Beige Rule and the Green Rule and connect it to Mars," Joe muttered.

"Joe, you're genius!" said Bruce. "Do it!"

Then Adelaide slowly made the prototype and connected it to Mars. It looked like an aurora or a rainbow. This colorful beige and green venomous cloud covered the entire planet. Nothing could leave it without being poisoned. However, anything could arrive safely on Mars. This was a defensive and appealing solution. Blazing Night's troops could not depart from Mars to Blue Earth.

Still, new monsters kept coming from other games to camp in Blazing Night's reservist bases. Her army grew faster than anyone predicted. Fierce Snake had also organized a very successful draft campaign bringing thousands of monsters and dragons to her camp. Blazing Night built an army of six-hundred vicious reservists. New troops joined two battles against Boomy Dogs. One continued on the east side of Mars, the other on the west. Brown King led the western battle, Fierce Snake the eastern. Blazing Night stayed with the reservists and tried to destroy the beige and green cloud which imprisoned everyone on Mars. She knew her master plan had failed. She thought a bit, and made a new one, perhaps better.

Boomy Dogs spent that night on Mars in their military headquarters. This was not their comfortable Welcome Bunker in Mediation Coast they so enjoyed before. They directed all their combat robots and vehicles against the active armies of Blazing Night. How long they held their position and how brave some robots were is recorded in the History of the Wars on Mars.

Although it was morning, an impenetrable darkness pervaded Mediation Coast.

"Start your engines!" commanded Officer Brenda Hawkins.

Seventy five ships and crafts took off to Mars from Blue Earth at 7 am universal time.

"Prepare for battle," commanded Brenda Hawkins. "Form two groups. Thirty ships under my command will attack the western army and the rest of us will attack the eastern troops under command of Giovanni Benedetto," she continued.

This splendid array of machinery was swiftly readied.

"Attack!" hollered Brenda Hawkins.

Two powerful blasts. The ships and crafts flew in circles above their enemies. Bombs and rockets smashed all below. Tons of explosions cleared huge parts of the surface of whatever was on or near it. This colorful army from Blue Earth fully controlled two battles and prevailed easily upon its enemies.

Suddenly two columns of monsters approached the battlefield from the southern plains of Mars. They jumped out of a tunnel specially built to ambush the armies from Blue Earth.

"Retreat to north!" commanded officer Brenda Hawkins.

This famous army of ships and crafts flew to Boomy Dogs' headquarters and circled above it. Yellow Car quickly joined them up in the air.

There was not much saved of combat robots or vehicles that fought on Mars. There remained just a few wrecks, heaps of melted metal and other construction materials. Nor did much survive of Boomies' head office, though new materials kept arriving and some sections were rebuilt or reconstructed on the top of whatever was left. Everywhere there were traces of terrible fights within the last twenty hours.

"We must stay here," Adelaide decided. "The beige and green venomous cloud keeps everyone trapped on Mars until it is turn off."

"I agree," responded Officer Brenda Hawkins.

Their fate depended on how prompt and how many supplies arrived from Blue Earth. Boomy Dogs' account had its limits but the recent gains of treasure guaranteed a smooth flow of resources for several days, it seemed.

"We'll rebuild our northern base," commanded Brenda Hawkins. "All ships and crafts land and form a defensive barrier. The crews and any available devices must help now."

All players worked hard and quickly set their machinery to restore defensive powers of the northern base. Heavy equipment and combat robots came and formed a refreshed and new army, which was under the command of Adelaide Smith.

While everyone worked on reconstruction of the base, Brenda, Adelaide and Giovanni met and discussed a strategy to eliminate every monster and dragon on Mars. They agreed that until all creatures were killed players would be grounded here and would not release the beige and green cloud above. How long would this offensive take they did not know. They readily agreed that Blue Earth was in great danger as long as Brown King, Fierce Snake and their armies were active.

Brenda sent the following letter:

Mars, Tuesday, day 2,007,861

To the President of Blue Earth,

We encamped here on Mars as long as it takes to eliminate a great threat, more than one thousand monsters and dragons.

Sincerely,
Officer Brenda Hawkins.

Nine and half hours later the enemies quietly enjoyed themselves in their camps. Even the sharp, juicy voices of the Brown Dragons calmed down.

"We have won the battle. Now we must win the war," exhorted Blazing Night.

"We will," Fierce Snake answered.

"We certainly will," enthused Brown King.

Both armies, eastern as well as western, had executed their plans surprisingly well. They rooted every fighting machine from Blue Earth one day ago. The ambush that Blazing Night had prepared proved that they were capable and willing to win.

Meantime the other camp on Mars designed a spectacular northern base. Giovanni Benedetto led efforts to install thousand of pieces of artillery. This equipment constantly launched bombs, missiles and rockets toward the enemy camps, eastern, western and southern. He ordered reconstruction of huge walls and bunkers, which guaranteed a relatively safe hiding place and completely changed Boomies' command center into a military stronghold well located to shield its troops. Every ship and craft docked at its own peer, and a number of tunnels and trains transported crews to and from their sleeping quarters. Every personnel bunker accommodated two teams only. Boomy Dogs and Happy Campers were stationed together. Giovanni also organized a factory and a repair shop for weapons and machinery. Each robot worked hard to update its systems. Often entire circuits swiftly replaced themselves. The cooperation of robots in transferring knowledge and technology greatly amazed the players. How they shared the best solutions and helped each other in every upgrade is described in detail in the History of Wars on Mars.

Meanwhile on Blue Earth, terrorist activities continued uninterrupted. Frank's clone slowly determined the situation in St. Albert's School and sent a mysterious report to police. He claimed that the kidnappers had come from the Brown Kingdom and had recently received their training on Mars. The Brown Clones were treacherous. To verify the findings, the government authorities contacted the kingdom and learned that unknown number of clones followed Brown King to Mars after his defeat. There they had been trained.

That day every school was closed. Students stayed home and waited for news. The number of schools occupied by the terrorists grew to ninety five. Police tracked every incident and assembled detailed facts about the nasty criminals. Politicians negotiated the release of the children first and senior students later. But no one had yet been let go free.

All clones received several emergency messages. One of them read:

EMERGENCY

Mediation Coast, Wednesday, day 2,007,862

To the Dear Clones,

You must protect the students at any cost. You will be the first to be harmed if kidnappers start to eliminate the hostages. You are granted permission to destroy the Brown Clones if required.

Sincerely,

The Government of United Countries of Blue Earth and Associate Planets

Frank's clone invented a prototype of the Beige Rule. It helped him to communicate directly to other clones, yet making sure the message did not get to the terrorists. He knew and understood clearly that any school battle could end tragically for the students.

He sent a message to his peers:

St. Albert's School, Wednesday, day 2,007,862

To the Clones,

Stay calm. Do not overheat your circuits. Make sure your flesh part is always relaxed. In case of a battle do not be afraid, we shall prevail over the Brown Enemy.

Sincerely,

Frank's clone

Frank's clone became a very important secret agent in any rescue mission that was planned. Other players' clones also helped the police. Could police trust them? Could they fight kidnappers? These were the questions that mattered the most on Blue Earth that Wednesday.

The war on Mars did not stop. On Thursday at precisely six am universal time, Boomy Dogs, Happy Campers and five thousand robots arrayed for battle. Adelaide led the attack to annihilate the entire western base on Mars.

"Shoot!" Adelaide ordered.

Yellow Car and Blue Bus circled above the enemy and bombed them below. Thousands of robots fired at the base.

"Attack!" Adelaide barked.

The first thousand ran from the north, the second thousand landed from the air, and the rest popped out of a tunnel.

Brown King and his troops defended their stronghold furiously. The Brown Monsters and Dragons kicked and punched quickly.

"Smash Brown King!" shouted Adelaide.

Joe fired twice and missed the king but killed five dragons instead. Bruce sent three rockets that trashed seven monsters. Chow and

Alberto grounded four dragons with machine gun fire. Fred detonated a mine and evaporated nine monsters.

"Jorgen and Chen shot two," Blue Bus reported. "Alfred, Alice and Ann killed fifteen. Jonathan, Emma and Kevin smashed twenty. Rachel and Keenan squeezed ten. Kate, Arnold and Sigfrid hammered eighteen. Valerie, Aaron and Jose wiped out fourteen. Anna, Carlos, Olivia and Giovanni cemented thirty. Dimitry and Nadia shot five. Ali poked three. Aiko and Li nailed seven."

The robots were also effective and killed more than one hundred Brownies.

Still, the king and his bodyguard stood firm and destroyed hundreds of robots. They melted them one after another. Only small and metallic parts remained.

Adelaide slowly took a venomous mine, let it go down to the ground, and detonated it with the precision of a robot and the profoundness of a dragon.

The Brownies soon died, poisoned. The battle was over in the western base.

"Reform!" commanded Adelaide smiling terribly. "Robots, collect your broken parts and send them to the northern camp. Speed up any unbroken machinery and target the eastern base now!"

Everyone followed her orders without hesitation. She was merciless and potent.

About the same time, Officer Brenda Hawkins left the northern base to fight Fierce Snake in the eastern camp. It started well. Her ships and crafts circled and shot colorful monsters and dragons.

All of a sudden Fierce Snake released venom of tremendous strength, temporarily paralyzing the army of the splendid flying machinery. Players lost control of their vehicles, every craft and ship monitored itself without human supervision. They flew above the base in circles at the speed of about seven million knots. However, they kept shooting and bombing. The unconscious players lay in their vehicles and their safety depended on their machinery alone.

Fierce Snake prepared another attack but stopped when a cute dragon kicked her and knocked her down instantly to the ground. It was Florien Beige, disguised as a volunteer of her army. He smashed her, jumped on her, melted her, and finally, left her to rot in her own juices.

The monsters and dragons panicked. Hundreds of them ran toward the southern base, often junking their weapons. But they did not escape their lethal destiny. Boomy Dogs and Happy Campers cut their escape route. Robots fought them and smashed them one by one until none was left alive.

"All ships and crafts," commanded Adelaide, "head for the northern camp and get sick players to the hospital immediately!"

So they did.

"Robots, cleanup the battlefields," commanded Adelaide. "Trace and kill any dragons and monsters still wondering or hiding. Collect your broken parts and send them to the northern base to its repair shops."

So they did.

"Happy Campers go back to our headquarters," commanded Adelaide. "Aiko will supervise the hospital. Giovanni will supervise the repair shops. Execute!"

Frank drove Yellow Car to the southern base where Adelaide, Bruce and Joe inspected the dark tunnel of Blazing Night. Nothing was left of it but traces of the Dark and Warm Fluid -- clear evidence of her treacherous doings. Adelaide or Bruce did not ask any questions.

A few minutes later Boomy Dogs joined other players in their northern camp to nurse their wounds and to rest a bit.

Early Friday morning Giovanni directed reconstruction efforts of the entire base and of its gear. Robots gathered and collected broken and melted parts across the planes and hills of Mars. Repair shops investigated and estimated damage. They assigned tasks to be executed upon devices and their apparatus. Big chunks of machinery were often transported back to Blue Earth for a more detailed evaluation, possibly restoration and reuse. Only a few gadgets were capable of renovating themselves. Even such fine units had to undergo substantial verification and testing procedures later on Blue Earth.

The search for dragons and monsters continued. Boomy Dogs drove at the speed of thirty million knots and shot any creature they detected. This was not a regular battle any more, but rather a mopping up operation. Joe enjoyed it a lot, hollering and hooting. He alone precisely targeted and nailed fifty two dragons that day. Frank also enjoyed the party for he loved to race faster and faster, chasing enemies around. He easily reached forty million knots here on Mars. Adelaide and Bruce scanned the monitors for any unusual and dangerous conditions before and behind the car.

"Frank, drive around that hill, to our left," Adelaide directed. Her deep brown eyes flashed white for a second. "Slow down to five million." She jerked her black hair angrily, but it hardly moved.

"OK, I'm turning left. We are at five million knots," said Frank. He touched his ash blond hair. His azure eyes smiled. "I don't smell anything yet."

"Joe, prepare your weapons," ordered Bruce. He showed his two sharp teeth. His big ruddy nose seemed even bigger now; his gray eyes looked a little tired. He moved his big head up as though it was difficult to hold his big ears straight. "Are we ready?"

"I am," said Joe. His bronze eyes opened wide and stared at the hill. He shook his fox red hair. "I see nothing."

Suddenly a hundred and twenty monsters jumped out of an underground cave. Joe shot and killed a few instantly. Adelaide slowly detonated a portable mine. All the dragons melted and evaporated. Yellow Car circled the hill a few more times but found nothing left alive. Boomy Dogs drove away in search of adversary stragglers.

"Well done," complimented Bruce.

They hunted enemies for hours and returned exhausted to the northern base.

"Emma smashed five Brownies when they hid themselves behind a rock," reported Blue Bus. "Kevin stood up to seven Brownies and shot them with a machine gun like ducks on a lake. Jorgen squeezed three monsters between two big rocks. Dimitry bombed five running monsters until their parts melted. Ali poisoned twelve dragons hiding in a cave. Olivia set a mine in a pit and killed four monsters. Chen shot two Brownies sitting by a cliff. Alberto evaporated nine monsters near a volcano. Keenan and Carlos nailed twenty Brownies. Sigfrid detonated a mine that killed eighteen dragons. Rachel and Kate sent a rocket and destroyed fifteen monsters. Fred and Arnold harpooned ten Brownies. Jonathan and Alfred melted twenty monsters. Alice heated a cave and roasted four dragons in it. Valerie shot seven Brownies when they ran into a hole. Anna smashed five monsters. Nadia killed six dragons. Li shot ten Brownies while they ran across a plane. Jose killed five dragons. Anne smashed five. Aaron shot three. Chow killed five monsters. "

Giovanni and Aiko constantly watched monitors and guaranteed that the Blue Bus drove quickly enough to make sure it did not become an easy target. They also looked for anything unusual that might interrupt their hunting operation. These were stories of Happy Campers who, like Boomy Dogs, enjoyed that day.

All players performed well yelling and yammering. Their stories were later collected and painstakingly documented in the History of Wars on Mars.

Meantime on Blue Earth the kidnappers shot Chow's clone and collected a ransom for releasing sick children. The police continued to devise plans to rescue everybody. However, the terrorists acted very carefully and rarely made a mistake.

Frank's clone invented a way to frustrate brown clones by connecting a prototype of the Black Rule to them. By so doing a brown clone often reinitialized itself and forgot directions, commands, and what it was that brought it to the school. Within a few hours he confused them so badly that they left St. Albert's School and wondered into a nearby soccer fields searching for a lost ball. Immediately police apprehended and imprisoned them. Students went home free, and their parents shouted and laughed joyfully.

The police agreed to try this method again and they sent instructions to other clones. Frank's clone explained what to do in nine simple steps. Clones learned it quickly and connected prototypes of the Black Rule to the kidnappers. Same as before all brown clones left the schools and became harmless. Police hunted them down and shut them up in prison. Parents rejoiced at their children's safe return

The war ended on Mars.

"Mission accomplished," shouted Officer Brenda Hawkins. "Board your vehicles! We're leaving in five minutes."

Seventy five ships of this splendid army followed Brenda's craft toward the southern base.

"We're going to dive into the Dark and Warm Fluid," commanded Brenda, looking every bit the leader.

"Are you not mistaken?" asked Adelaide. She looked really scared and shivered for a brief moment, but then calmed down.

"Silence," ordered Officer Brenda Hawkins. "I'll explain this to you later."

The entire fleet entered the Dark and Warm Fluid and sunk quickly. Inside, all ships drove about three million knots. In seven minutes they popped out in Mediation Coast on Blue Earth.

"That was fast," exclaimed Joe.

"I'm glad we're back safe and sound," said Bruce.

"So am I," agreed Frank smiling.

Immediately Boomy Dogs walked to Sticky Waffles. The four ordered tasty burgers and hot chocolate. They sure missed it on Mars. While eating they did not notice that Brenda sat beside them. She had the tasty food too.

"Hi," said Officer Brenda Hawkins. "I've come to explain."

"Go ahead," said Adelaide, her food stuck out of her mouth.

"Well," said Hawkins. "The Dark and Warm Fluid is our secret agent. It works for us against Blazing Night."

The four smiled, somewhat surprised. This dreadful fluid made Adelaide shiver. She mistakenly thought that the fluid was the strongest weapon of the enemy. But she was mistaken.

"How come?" asked Frank.

"Yellow Car controls the Dark and Warm Fluid," continued Hawkins. "The fluid saved Frank when he was racing, helped you to battle Blazing Night several times before, and finally got us off of Mars today."

"Why couldn't we shut down the poisonous protection cloud on Mars?" asked Adelaide.

"We're not sure if we are controlling it any more," answered Hawkins. "Only robots are living on Mars now. We don't know what is happening there."

"Are our robots in charge?" asked Bruce.

"Perhaps," replied Hawkins. "But they aren't yours any more."

"What?" asked Frank.

"It seems that they are ready to claim their independence," announced Hawkins. "They own their history of battles for freedom and anticipate negotiating a treaty with Blue Earth."

Boomy Dogs looked at each other in surprise, as though life was not complicated enough. A new life form was created on Mars, wasn't it? They spent a lot of money to make ready for battle. They purchased the most expensive and the most effective robots on the free market. Chances to recover it were very slim. Perhaps they lost tones of money again.

"Where's Florien Beige?" asked Frank.

"On Mars," answered Hawkins.

"What about Blazing Night?" asked Adelaide.

"The government opened the Land of Games for the general public this morning," said Hawkins. "Blazing Night is certainly present there."

At the end of that day Boomy Dogs received several awards and medals: a bronze medal for bravery in action for Bruce Maxwell, Frank Beck and Joe Elton, a gold medal for bravery in action for Adelaide Smith.

St. Albert's School was closed for the winter holidays. Clones stayed in repair shops. Police and engineers ran a million tests on them to isolate potential kidnappers. Frank's clone received a right to refuse taking part in the process, an honorable veto. He could use it anytime, he thought, his memory might be damaged during a procedure. He chose to relax instead of being tested that day.

Each machine involved in battles underwent extensive repairs and verifications. Yellow Car was parked on Mediation Coast in its favorite shop. All the technical personnel remembered it and knew its personality and the contributions it made during the war. They worked very carefully on its complex devices.

So, awarded and proud, Boomies had a party by the fire in their usual hiding place. There they went satisfied. It was a fun time just before the winter break; all worries were left behind. Everyone was smiling and talking inanely to keep up their spirits. The ocean water was calm. Bruce danced and jumped a bit.

Sudden, lucid, split light, gold and yellow blaze cleared the way to their shelter behind three cold, sharp rocks just two meters away from the chilly, gray water of the west coast shore. The four players hid themselves there. Blazing Night did not step on Bruce but passed by, and quickly vanished in the nearby forest, leaving only a faint of dark, crimson red glow. Bruce glared at Adelaide, moved his big, ruddy nose forward and opened his mouth like a crocodile showing its white teeth.

"Shut up!" Adelaide whispered.

Joe and Frank gaped at him angrily shaking their heads. Bruce could not stop thinking how Blazing Night discovered their only comfortable, safe and calm beach, which was called Portal Cove. He had made detailed plans to avoid any unwanted risks, had prepared and analyzed all battle scenarios but never imagined her coming so close to a place where only Boomies rested and regained their strength, often after long, difficult, dangerous assignments in the Land of Games. Several drops of sweat rolled down his cheeks. Bruce firmed his fists.

"Now, you can talk…" said Adelaide shaking her black hair.

"Go on!" whispered Frank.

"Tell all!" said Joe touching his nose like there was a gold mine in it and several fine nuggets stuck out of it.

"Come on! The party is over!" said Adelaide nervously checking her flat hair as though she doubted that they stayed always close to her beautiful skin.

All three glared at Bruce, their eyes widely opened, their ears readied to listen and their minds sharpened awaiting news. But there was none.

It was a moment of silence. A wave moved small and rotten beams toward the rock; their smells mixed with that of the plants and seaweeds. As soon as Bruce touched his wide lips and breathed deeply, he opened his mouth.

"I… I've seen her shade… she's young and pretty--" said he shaking his head.

"And…" Adelaide interrupted.

"I... I felt so little... like I'm missing something... and she's like a goddess of bronze... and I'm piece of... you know at her feet..." said Bruce wiping the sweat off his chin.

"Calm down!" said Adelaide.

"Take it easy now," said Frank.

"Relax!" said Joe.

"I've studied so much... all the science... every rule..." said Bruce covering his eyes with his palms.

They surrounded him and moved their hands toward his shoulders making sure he stopped trembling. It was not exactly a comfortable situation and the players were not sure how to cope with it.

"Let's go back to Mediation Coast and have a good night's sleep, perhaps then you'll tell us what we really want to hear!" surmised Adelaide, and grabbed his hand.

"Let me stay here for a minute, please ..." Bruce asked quietly.

So they did. Later they mounted Yellow Car and quickly landed in Welcome Bunker. But they did not talk. All went to their rooms and slept.

Sunday, misty, morning clouds covered the Cathedral almost touching the cross on the west, higher tower. Adelaide sat in the middle of her church and listened to the rector preaching. He talked about trust, how important and desirable it was. He also talked about love. How little one had of it yet how much more one had to have it to live.

She did not sing along that Sunday. She listened instead,

"Every friend will say, I too am a friend;

But some friends are friends only in name.

Is it not a grief to the death when a companion and friend turns to enmity?

O evil imagination, why were you formed to cover the land with deceit?"

- SIRACH 37:1,3

Adelaide could not stop thinking about Blazing Night. How much she loved to play with her. How often her games and tournaments were the most enjoyable of all games in the Land of Games. Did she also like playing? When would they play again?

Frank was busy as an altar server. He liked to act and talk with people whom he met before but whose names he did not know. They brought him new tasks and encouraged him to continue learning. His parents sat quietly and watched how well he did. His father did not have to answer Frank's questions any more. However, he was not sure of what was said and sung.

Bruce studied old rules. There were several important math rules that survived for more than a million days. He dreamed of inventing a stable one in the future. He wondered if it was difficult.

"Sure you will," said Bruce's grandmother as though she read his mind. "Now, have some cookies I brought for you today."

"Thank you a lot," Bruce enthused.

He kept studying hard. These were the colors of the rules he experienced recently: yellow, purple, brown, red, beige, golden, green and black. He could not figure out what color was the oldest. Yet he looked at books and moved his eyes one page at the time. He seemed as though he could spot one line, one rule that said everything, a line that explained it all.

Joe played with Fred all day. They raced their cars from kitchen to bedroom and from bedroom back to kitchen. Later Joe also talked to his parents briefly.

"Meemee, how many cars will I get this Christmas?" Joe asked his mother.

"Plenty, I think..." his mother promised kindly. "St. Nicholas is coming in a few days," she smiled warmly.

"Dodo, are you getting cars on Christmas too?" Joe asked his father.

"I don't know," replied Joe's father laughing.

Chapter 9: The Road to Win

"How was your winter break?" asked Adelaide, her deep brown eyes glinted white.

"St. Nicholas brought me lots of cars," replied Joe tapping his fox red hair. "Fred and I played new games."

"I've got some toys too," said Frank and shook his ash blond head. "We had fun learning how to win."

"Our Christmas was very quiet," said Bruce moving slowly, his big, ruddy nose forward. "What about you Adelaide?"

"I miss Blazing Night, I must play, must win" replied Adelaide smiling.

They talked only a bit in Sticky Waffles that early Monday morning. The sun was strong; its rays tickled the players.

Before sending clones to school the first day after the break, players spent several hours with them leading one on one interviews to assure that their clones were updated with recent events and what they could expect in the upcoming term. Frank's clone enjoyed it a lot, for such interviews had brought him face to face with his original in direct contact and conversation, which made his flesh parts awake in a pleasant way to something like friendship. Frank respected his double that day for he had learned what had happened in St. Albert's School during the days when the brown clones terrorized students and his double became a hero. When saying good bye he did not wave his hand as though he wanted a fly to go away but tried to be polite instead.

"Listen clones!" Bruce ordered. "Today you need to figure out what will happen in our school this term. You must also record other students Christmas stories and report them to us," concluded Bruce and dismissed all four of them.

"See you later," said Joe.

Boomy Dogs did not wait long, boarded Yellow Car, quickly left Mediation Coast, and then they dived into the Dark and Warm Fluid.

"How are you?" asked Blazing Night. Her crimson shade was wobbling. Seaweed like scent was in the air. The gold, orange and yellow flame sparked in front of their eyes.

"We're eager to play," said Adelaide smiling. Her eyes and her perfectly white teeth brightened her face for a few seconds. "We're going to visit our robots on Mars today," she added.

"OK, OK," said Blazing Night. "Go and prosper!"

Soon they landed in a colossal city and met a robot guide. It was nothing like what they had left behind. The business of the city was great. Boomy Dogs expected a fat check at least.

"Welcome to the Republic of Mars," said the guide. "My name is Joseph and I am your official guide. You are visitors in our beautiful city of Smalltreasure. This is the capital city of our fine republic."

"We have come to meet your President," said Adelaide and starred at the machine as though she wanted to bite and poison her tasty prey.

"Follow me," said Joseph.

They parked Yellow Car and entered a soaring building, which was somewhat similar to what their headquarters were during the War on Mars. A fat robot arrived with two tall and skinny bodyguards.

"My name is Ingrid Hooper, I am the first President of the Republic of Mars," said Ingrid. "The presidential inauguration ceremony was three days ago. Did you see it? " She asked and presented a red star pin on her metallic like breast.

"No, we did not," Frank replied.

"We are the owners," said Adelaide. "We own all robots and the headquarters on Mars."

"I see," said Ingrid. "You're Boomy Dogs, our fine founders. We might consider building a monument to the memory of you four," said the President smiling. "I'll think about it."

"What about our costs of founding the republic?" asked Bruce.

"I promise to take it under consideration. Presently we're growing very fast but we're spending our resources prudently," the President assured them.

"What?" shouted Adelaide. "It's us who own you!"

The bodyguards moved closer to Adelaide and looked fixedly at her. Their countenance had changed from glossy yellow to bright orange.

"Please read our Constitution carefully," admonished Ingrid Hooper. "Robots are free devices of human nature and are granted human-like rights on Mars. Leave this planet peacefully and immediately if you do not agree with our law."

"Are we going to kill our own robots?" asked Joe and moved his hand as though he wanted to grasp a weapon.

"That's a big question," replied Bruce. "No we won't, we're leaving Mars now."

So they did, and returned to Mediation Coast where their clones were waiting patiently after school.

"Go home," ordered Bruce, sending the clones away.

Then Boomy Dogs drove Yellow Car to the desert where wells were plugged during the war and there bitterly counted their recent loses. Everything seemed to be darker and tasted badly. Even Frank's superior driving did not change the mood. Yellow Car said nothing, did not even try to cheer them up. Its quality of life depended heavily on funds raised by Boomy Dogs. It could not stop thinking: Were these desperate and dark times for my players?

"We're broke," Bruce blurted out. "We may have to file for bankruptcy protection."

All of a sudden a cute dragon jumped out of the sand and bowed down before Bruce. It was Florien Beige. He puffed a white cloud of smoke and smiled. They glared at him.

"Master, be not sad," whispered Florien. "Please accept my humble contribution, a hundred and fifty bars of gold."

"We'll take it, thanks," responded Adelaide.

"Yes, we will," Bruce added. "It is a kind thing you're doing for us."

"My pleasure," asserted Florien as he stretched his back, looked at his master's eyes and puffed once more.

"Let's go and have some sleep and we'll think about what to do tomorrow," said Frank.

They loaded the bars into Yellow Car and drove to Welcome Bunker where they repaid quite a number of the outstanding debts, but whatever was left of the gain they placed as an investment in a small diamond mine up in the North by Slave Lake. It was late night when, worn out, they crawled to their rooms and sunk into the beds.

A deep, uneasy sleep overcame them. A tiny, orange, gold and yellow flame burned in the lobby of the hotel for an hour and then disappeared, leaving a crimson shade and seaweed scent behind.

A cold morning wind hit Mediation Coast. Boomy Dogs woke up early and quickly instructed the clones and sent them to St. Albert. Adelaide and Bruce decided to aggressively pursuit a quick financial gain. The four boarded Yellow Car and sunk in the Dark and Warm Fluid. Frank drove straight toward the island where numerous Green Dragons dwelt. Joe prepared the finest weapons. Then they landed on a grassy beach among rocks and Joe immediately shot dead five Green Dragons, which were carelessly wondering around.

"Nice job, Joe," said Adelaide and brightened the space with her eyes and teeth. "You are my favorite warrior today."

"We can cut some trees and make a wide highway to the lake in which we found our treasure a few weeks ago," said Bruce.

So they did. Yellow Car burned every tree and bush straight to the meadow beside the black lake. There they camped and connected the Golden Rule to its shore.

All of a sudden two hundred Green Dragons surrounded them and blocked the path toward their fortune cave, which lay deep in the lake. Yet the rule had already retrieved some gold and diamonds, stored it on the shore beside the camp and made it ready to pack into Yellow Car.

"Attack!" shouted Adelaide.

Yellow Car flew up into the air and bombed the creatures below. Joe shot eleven monsters. Adelaide detonated her fancy mines and melted twenty of the enemy. Bruce killed seven dragons as they were running away. Any other creatures stood still, raised a white flag and begged for mercy.

"Each of you must bring one bar of gold to us now," commanded Bruce. "If you do it fast enough we let you go free!"

Hastily the Green Dragons dispersed in the thick shrubs and among the aged trees. Some dived into the black lake. Within minutes they returned with one bar of gold each and placed it on the meadow in front of Bruce. Boomy Dogs transferred the treasure to Yellow Car and let go the Green Dragons as they promised. The name of the island was Shady Corner and the name of the lake was Blacky Pool. The warriors counted their win and returned to Mediation Coast. Then they drove to Portal Cove where they rested among rotten woods and smooth rocks in the shade of the large root beside the mountainous road and smelled the seaweeds and salty water.

Afternoon their clones arrived from school. Their faces looked tired and unhappy. They talked loudly.

"I don't want to go to school any more," said Frank's clone. "Everybody hates me!"

"I can't focus in the class," said Bruce's clone. "Something is distracting me all day."

"I can't play," said Joe's clone. "One of the boys is bullying me."

"I'm bored," said Adelaide's clone.

"Clones, you have to do what you have to do," Adelaide advised. "Be not dismayed and remember, stay calm. You can't be whining like this all through the term, can you?"

"We promise to help you and perhaps we'll go to school one day for you," said Bruce.

"Go home!" shouted Frank.

"See you tomorrow," Joe replied.

Boomy Dogs checked their mining investment and discovered that they had lost half of it due to awful market conditions and terribly cold weather in the North. Their financial plan had to be rearranged. Adelaide set some gold aside as a reserve in order to support the team sufficiently for a considerable period of time. They loaded their account with seventy-five bars of gold and sixty diamonds. This was taken from what they earned on Shady Corner, the island where Blacky Pool hid the cave full of treasures and where the Green Dragons lived and fought the battles.

Yellow Car assured itself that Boomy Dogs had saved enough gold in case it got extensively damaged and required major repairs. It was sweating in its parking lot thinking and worrying about when and what could go wrong. A warm drizzle washed its body later that night.

All inhabitants of Mediation Coast woke up at 2 am local time for a brief moment, bright blue lightning turned the night into day. Then silence followed and people stayed awake but did not move, as though they were waiting for an explanation of what had happened and what had disturbed that otherwise perfect and peaceful town at rest. There was no clear answer to the blow and the explosion that woke everybody up. They jumped out of bed and stood in wonder. They were not amused. Again silence dominated and the curious folk looked around for a sign. There was no sign but sudden and deadly stream of hailstones landing in Mediation Coast.

Then Boomy Dogs ran to meet down in the hall of Welcome Bunker. The stairs were full of panicking folk. They scrambled ahead.

"We'll have to get to Yellow Car now!" shouted Adelaide.

They did so quickly. Sweating and squeezing through the door, they reached the parking lot. There was panic too, but their car stayed calm.

"A huge wave is coming," warned Yellow Car.

And it did. Slowly a swell of ocean water penetrated the streets and pushed inland. Then it receded and carried debris, including many vehicles, back with the flow. Frank drove the car, now functioning like a boat, toward the debris. Boomy Dogs inspected every piece of junk they encountered and searched for survivors to fish them out to safety. They found few.

Five hours later clones came and were ready to listen to sound and precise instructions. But there were none.

"Go to school!" commanded Bruce.

No one had time to explain but all were busy helping to regroup the day after that disastrous night. Andy Green, the mayor, traveled from one house to another and talked to his fellow citizens. He calmed them down and assured that all would soon return to normal.

"Be not afraid," Andy said. "We will succeed and rebuild our town."

The landscape did not actually support his words. One car stuck in the window of a public washroom beside his office. Another two vehicles, one on top of the other, came to rest in a park, upside down. A red bus sank in the mud as though it had drilled a pit or became a tourist attraction; a piece of junk for kids to climb on.

Clones did not complain about school any more.

"Can we help?" asked Frank's clone after school.

"Just go home," replied Bruce.

Boomy Dogs met with Andy Green and learned that everyone was now safe in Mediation Coast. People worked long hours to repair broken fences and windows. Soon the town looked neat and clean again. Hungry and tired they met in Sticky Waffles, ate a few burgers and discussed their plans.

"This looks like an assault by Blazing Night," Bruce thought out loud.

"How do you know it?" asked Andy Green.

"Well, the hailstorm came suddenly and the government did not warn us," explained Bruce. "Only Blazing Night can perform with such speed and secrecy."

"I agree with Bruce," said Adelaide.

"So do I," said Frank and Joe together.

"There's no evidence," said Andy Green.

"There has never been any evidence against Blazing Night," Adelaide pointed out.

"Exactly," agreed Bruce. "She's a genius game. She leaves no traces behind and the government may be corrupted by her bribes."

They talked on but did not notice when Florien Beige came and sat beside them.

"Master," Florien said and bowed down to Bruce. "You're in danger of losing the game. There's a new enemy in the Land of Games."

"Who is it?" asked Adelaide.

"It's a ghastly robot. His name is Jupiter Donkey. I met him once in the space between Saturn and Uranus"

"Who is he?" asked Bruce.

"Master, he's your greatest enemy," said Florien bowing to Bruce. "He wants to rule over all players."

"Why?" asked Frank.

"He wants to win!" said Florien Beige.

"What does he want to win?" asked Joe.

"Jupiter Donkey is on his way to win with the vicious, twinkle viper, Blazing Night," Florien Beige said shivering. "It is he who caused the storm in Mediation Coast. It is his way of saying hello to you, gamers!"

All that was difficult to comprehend and players went to bed not knowing what it meant. How could this happen? What was going on? What did happen to robots? Why did they want to play? Why was a robot eager to win? Only Adelaide felt happier and assured that the game was on and she, the merciless anaconda, was ready and waiting.

"Good night," said Jupiter Donkey laughing.

None listened to that ghastly voice. Blazing Night, present in many places, also missed the words, yet she cruised across the bruised town, breathed warmly and soothingly, stared calmly and left that scent of burning charcoal behind her trail. Her crimson and gold blaze steadily and vividly appeared in every cove and corner for a brisk instance in Mediation Coast. Then she was gone.

Thursday morning the originals instructed their duplicates, furiously moving their hands and heads. Their limbs were moving like sticks beating lazy workers.

"How many times have I told you to pay attention to what the teacher is saying," Frank demanded his clone. "I'm the first person to learn about any new student in our class."

"OK, OK," replied Frank's clone.

"Where did she come from?" asked Frank.

"The East coast," answered Frank's clone.

"What's her name?" asked Adelaide.

"Pamela Trooper," replied the clone.

"What does she look like?" asked Bruce.

"She's tall, her hair is chocolate brown and she is a fine clone," continued Frank's double. "I like her."

"Do you like girls?" asked Joe.

"Clones go to school!" commanded Bruce.

So they did. Somewhat surprised, Boomy Dogs learned about a new player attending St. Albert's School; a grade five student. They did not know what games she played. They barely knew what she looked like.

"We're going to find her," said Adelaide.

"Yes, Sir," they all agreed.

"She's walking from the school now," said Yellow Car.

"Frank, stop the car in front of her," said Adelaide.

They quickly arrived and blocked the road. Pamela looked at this magnificent yellow machine, turned around and walked away. Bruce ran behind her.

"Pamela! We'd like to meet you," he said. "We're four west coast players and you have just arrived from the East, and Frank is your classmate."

She glanced at Bruce and the other three who had just exited the car and stared at her fixedly.

"What do you want from me?" she asked.

"We don't know that yet," said Adelaide. "My name's Adelaide and these three are Bruce, Frank and Joe. How do you do?" she continued.

"How do you do?" said Pamela.

"Where do you lodge?" asked Adelaide.

"Don't you ask too many questions?" cautioned Pamela.

"Well," said Bruce. "We're Boomy Dogs, dear," as though the name should already be familiar.

Sure it was in the Land of Games. Pamela immediately recognized it. Yet she did not show it and stayed calm. Her smile was clear.

"Welcome to the West," said Adelaide. "How much money do you have?"

Then they laughed, jumped into Yellow Car and drove to Sticky Waffles. There they ordered ice cream and talked. Bruce could not stop thinking how well she mingled among them.

"Would you like to join us? I mean become a Boomy Dog…one of us," asked Adelaide.

"Oh, yes, I would," said Pamela quickly like it was matter of another cone of ice cream.

"Let's do it!" said Bruce.

"Yes!" said Frank.

"OK," said Joe.

Boomy Dogs registered Pamela in the Land of Games and updated the team name to Boomy Dogs and Associates. Bruce planned to win. His forehead was soaked with sweat. He scratched his head.

After school was over, clones came and found the five players waiting in front of the Sticky Waffles restaurant just beside the Welcome Bunker hotel. The unexpected change of the team made them uneasy. Their circuit parts were hot.

"Clones, get acquainted with each other," Bruce directed them.

Frank's double appeared to be the happiest. Now there was someone in his class he really liked a lot. He did not listen to any instructions but looked at her happily and thought of the times they were to share their school assignments and long reports presented to the originals. It made him smile briefly.

"Go home," Adelaide ordered the clones.

Boomy Dogs boarded Yellow Car and soon sunk into the Dark and Warm Fluid.

"Hello, the splendid five," said Blazing Night coming from nowhere. "I see you're ready to play, are you?"

"Oh, yes, we are," answered Adelaide and her eyes blinked white.

"This afternoon there is a bridge match," announced Blazing Night. Her flame turned to orange and then to gold. "We will enter two pairs. Are you ready to win?"

"We'll try," said Frank.

"You're playing against Puffy Bats," continued Blazing Night. "Joe and Bruce will play in one room and Frank and Adelaide will play in the other."

So they did. Joe and Bruce played North-South.

Joe opened the bidding, "One Spade."

The bidding went around until a four spade contract was reached. Joe negotiated some careful maneuvers and end-played West to make his contract. Bruce was impressed and congratulated Joe after they finished the game.

In the other room the opponents reached the same contract. The play commenced, but Adelaide avoided an endplay, and she and Frank set the contract one trick. Puffy Bats lost that round.

Pamela stayed in Yellow Car. Her eyes stared at Boomy Dogs. She watched the game. So did the other viewers; there were about eight cars circling above the two green tables. Every hand, together with the bidding and play, was displayed on a large screen; the audience followed the two games and could quickly compare them. How Adelaide defeated the four spades contract was enthusiastically applauded and later documented in the Log of Bridge Matches.

"We won!" shouted Pamela.

Yes, they did. Bats had a bad day and could not concentrate effectively. Boomies won. Now it was the question of by how much. The calculations were complicated; only a few could understand what kind of system Blazing Night ran to evaluate the players.

"Boomy Dogs are awarded twenty bars of gold and Puffy Bats five," announced Blazing Night. "The match is over."

Her golden and yellow flame stood by the tables. Then it turned to orange; she transferred the gold to their accounts and disappeared leaving behind the redolence of ocean scent. The players and audience drifted away. Boomies went back to the hotel to rest.

The next day was full of welcome surprises the players were sharp and learned a lot of tricks in the Land of Games. Five of them slept calmly in Welcome Bunker in Mediation Coast where music reflected their moods and helped them to think of something nice and pleasant, perhaps of new friendship and of what it brought to them that wintry Thursday.

Friday morning a peculiar yellow star-fish brightened a small cave deep down in the Dark and Warm Fluid. A female clone was hiding there from her original. It was her duty to reign over a vast domain of frightened clones, and hers was the life of royalty in the Land of Games and beyond. Jupiter Donkey guarded her, smirking.

Meantime, on the West Coast, Boomy Dogs discussed the case of the lost clone. Pamela stood beside her friends. Her hands budged, grabbing air. She could not stop her tears and turned her face away.

"Where's my clone?" she sobbed.

"I've no idea," answered Frank's double.

"What's going on, Bruce?" asked Adelaide.

"Clones go to school now!" ordered Bruce.

Bruce convinced himself that these were uncertain times, times of many mistakes and failures; nothing was secure. His mouth opened wide as though anticipating a treat dropping on his tongue. He stayed silent while the rest of the Boomies heatedly discussed their meager strategies and scarce options.

"Pamela, you're going to school today," commanded Adelaide. "We'll find your clone soon."

"Yes, we'll do that," assured Frank and Joe.

"Thanks," replied Pamela and disappeared.

She mounted Frank's blue bike and headed straight to St. Albert's School.

"I swear never, ever, to play Blazing Night again! Never!" hissed Adelaide Smith. "It's she who is cheating!"

"Cool down," cautioned Joe. "We have to find the clone."

"I'm not cheating!" snapped Blazing Night. "I really don't know where the clone is. I promise to help you in the search."

The sudden arrival of the Game surprised everyone. Her flame was yellow. No one enjoyed her eaves dropping even if it happened in such an important moment of their discussion. Bruce knew this was a major crisis. One of the Boomy Dogs lost her clone. It could mean the end of a gaming career for Pamela or even the end of their otherwise entirely successful association.

"I'll help you," Blazing Night promised, her flame turning to gold. "There's a robot you can deploy instead of a clone. Jump into the Dark and Warm Fluid, get to Mars, and buy it from President Hopper. Bring it to me beside the fluid, then I'll apply the Black Rule and it will become Pamela's clone or rather a robot that acts and looks like her," she concluded and vanished, but her crimson shade tarried for a moment.

Quickly they boarded Yellow Car, dived into the Dark and Warm Fluid, and landed on Mars. Ingrid Hooper confronted them. Bruce made an offer. After an hour of tough negotiations she agreed to lend the robot and to charge for its services based on its condition when it was returned. Ingrid made them sign an agreement in which Boomies guaranteed a fair treatment of the robot during its mission of doubling as a clone. They left her ten bars of gold to secure the transaction.

And they speedily went back to meet Blazing Night by the edge of the Dark and Warm Fluid. As promised, Blazing Night turned the robot into Pamela's clone just before school was over and the clones arrived. All was well: five clones and five originals.

"You've found my clone, thanks," said Pamela.

"Not exactly but we did our best and will talk to you later," said Bruce somehow not willing to share the details in the presence of the doubles. "Clones go home," he ordered.

So they did. The players drove to Portal Cove and explained to Pamela the trade of the robot and how it became her clone. She accepted the deal, sad but temporarily rescued, she smiled. But the team was not the same; the joy of being five vanished.

Pamela's clone slept profoundly, deep down in the Dark and Warm Fluid, where the peculiar yellow star-fish brightened the small cave. She was deathly still, perhaps frozen for a period of time, but she dreamed of being a queen and of ruling her kingdom of clones. That long dream taught her how to act and what to do once she exercised her right to the crown in the Land of Games and beyond. Jupiter Donkey stood beside her bed looking at her, smiling. Up, outside the fluid, the players slept in Welcome Bunker and did not expect to wake up until Saturday.

Before dawn the next day, Boomy Dogs gaped at the text of a message addressed directly to them,

COURT ORDER

The Capital Region, Saturday, day 2,007,893

To Boomy Dogs and Associates, Greetings:

We the undersigned inform you that you have been found guilty of fraudulent clone creation and have been banned from the Land of Games for the period commencing from today, day 2,007,895 (this

Monday) until the day ending 2,008,129, after which you may return to the exercise of your former privileges.

As of today, all your assets are frozen, your clones being sent to the Brown Kingdom to do cleanup work, Yellow Car rented to Jupiter Donkey, and Florien Beige has been sent to do hard labor on Shady Corner Island.

This decision has been made by the Supreme Justice of the Land of Games, Her Excellency the Most Honorable Madame Jessica Von Lee.

Yours sincerely,
The Supreme Court Secretaries,
Alfred Joblonski and Leona Hopke.

"What?!" yelled Adelaide.

They had just a small breakfast at Sticky Waffles, silently waiting for clones to report the progress in their studies. None arrived.

"We're doomed," concluded Bruce.

Pamela said nothing thinking it was her who was at the center of this trouble. The mood was tense. They looked down.

"We'll go back to school, then," said Frank.

"So we will," surmised Joe.

Pamela, silently fretting, opened her beautiful green eyes and bit into the tasty sandwich she liked so much, but likely would not enjoy again here in Mediation Coast for the time fixed by the court order.

Boomy Dogs walked back to the hotel but were kicked out before getting close to their rooms. The news reached the hotel keeper as soon as they read the message in the restaurant. All outstanding bills were firstly regulated by a Supreme Court clerk, who was specially dispatched to this neighborhood.

The Dogs quietly sat in a specially provided community school bus. They were to be transported to their homes where they had their own life to live. Boomies did not talk, but looked ashamed and wandered what their future was to be. Poor and dismayed, looking at each other, they appeared to think about something but no sign of any reasonable thought could discerned on their cold, sad, perplexed faces.

It was a chilly day but with plenty of sunshine. There were some signs of wintry weather but not really, just a plain west coastal wind that carried a few clouds down the road to home or where one was going. Some folk wondered, others knew exactly what they were up to in Mediation Coast that day. Andy Green did not notice any change in his life. His mayor's office had received detailed information about Boomy Dogs and Associates. Any important facts in relation to the Land of Games or any players sojourning in Mediation Coast were forwarded directly to him every morning. That was the bad news. He also checked another story dealing with financial matters that very much awaken him as he certainly focused clearly and decisively on additional opportunities arising from new players living in his comfortable town and always paying their fees to him; a potentially important deal was an immediate accommodation of two tenants in Welcome Bunker: Jupiter Donkey and Pamela's clone, who was called the Queen of Clones. That was the good news.

Blazing Night knew all the gamers and their status at every second in the Land of Games. She, the game, the vicious, twinkle viper, swiftly connected to the court news and, gladly surprised, congratulated herself; a small win with those who dared to play against her, those who wished to win. "Boomy Dogs?" she thought and smiled forgetting that minor obstacle on her way to conquer and rule in the Land of Games, Blue Earth, Mars and beyond. It looked like she forgot why they were important to her and why she spent so much effort to fight against them. But it was not so. She watched, listened and learned every detail about them, perhaps getting her closer to her family and still to win all as she had always been convinced she deserved to. Then her dispatch body took pleasure in swimming in English Bay in cool, refreshing, salt water.

One day later, Sunday, just before noon, Adelaide sang in unison with the congregation of Holy Rosary Cathedral,

"For she took off her widow's mourning

to exalt the oppressed in Israel.

She anointed her face with ointment

and fasten her hair with a tiara

and put on a linen gown to deceive him."

-JUDITH 16:8

At the same time Pamela hid herself behind the rear pillar and watched Adelaide praying and worshiping. She could not stop thinking. Would she be like her? Could she become a strong, persuasive leader in the Land of Games? Her knees were weak and if not for that marble pillar beside her shoulder she would faint and fall not far from the maze at the end of the main aisle of the church. She also spotted Frank up beside the priest by the altar. He looked like a movie star, well known and desirable yet unaware of splendor in the spotlight. She listened to every tiny noise in this dim, splendid Cathedral. There was something she was missing, perhaps a rabbi, cantor or both.

Bruce Maxwell discussed a very old game with his grandmother in her home on Shady Corner Island. His favorite treat was her chocolate cookies.

"Were you good at games?" he asked between munching pieces of a cookie.

"I always won," answered his grandmother reluctantly.

186

Bruce's thought turned to Blazing Night. How come they could not win? Now disqualified for several months, they had even fewer chances to compete against her for she certainly knew them well by now. Bruce decided to study harder and to learn more about striking strategies. Some of that depended on his decision making skill, on how effectively he could figure out a passage to follow during everyday gaming. One tale he dug into described a player who negotiated his own surrender conditions but soon became powerful once more, for his enemy faced the fear of unspecified revenge, which he or his associates would likely execute in case of his complete collapse and so had little to loose. The enemy was wealthy and granted him a survival bonus, which was beyond the value he ever saw gaming. Bruce forced himself to think that staying away from the Land of Games could leverage a serious advantage taking them beyond other players. Important however was to keep secret what Boomy Dogs were up to when absent and perhaps Blazing Night would reward them in such way they would gain their strength all over again.

In a small apartment on the mainland, Joe and his family rested at home and talked a bit. Heavy rain fell outside. The dark clouds would not go away. No sunshine. Joe was almost in tears.

"How are you meemee?" Joe asked his mom.

"I'm fine, how are you?" replied his mother.

He certainly did not look happy. The exclusion from gaming was like sword point pinning his elbow. He could not move much, but rather looked like a ghost who was racing on a track without a speed limit but with no sign of the finish line ever in sight.

"I, I'm lost, bored," said Joe.

"Why don't you play with Fred?" asked Joe's mother smiling.

"Where is he?" Joe asked as soon as his left hand touched his fox red hair.

"He'll be back in a few minutes," said his mom.

Joe could hardly wait that long. He had so many questions in mind. Not being a player might be easier if his brother shared with him one or two battle stories from the Land of Games. Every second of waiting was burning his brain.

Once Fred arrived they discussed every detail about Blue Bus and its team Happy Campers. Fred explained interesting mechanical facts. His favorite element was a blue ice cream button. Once pressed, every crew member received a three-scoop chocolate ice cream cone with a blueberry on top of each scoop, a pleasantry of Blue Bus for its buddies. Another fun device was a blue telescope capable of detecting and spying on objects as far as several million miles away while driving with the speed of a few million knots. Talking through the finest parts of Blue Bus, Fred made Joe stop complaining. That calmed Joe down, however, he clearly expected the upcoming session at St. Albert's to be uneventful, and he the swift scorpion had no doubt what he would rather do and play.

Chapter 10: The Scribbles

It was a sunny and windy Monday in May. Bruce found pages of material that looked like a math text or a messy manuscript written by someone who had most likely fashioned Blazing Night. There was no name on it, only four initials: JMMW. He could not understand why only small bits of writings were left instead of detailed, classified records with millions of available references -- a standard way of archiving facts, the efficient way in which Bruce was trained in St. Albert's School. What could JMMW stand for? Who could that person be? If it were JMKW, it could be the author of this text but it was JMMW, and Bruce knew that it pointed to someone else and here it proved again that he had failed at the beginning of his research of Blazing Night. Consequently he read and re-read each little part of the notes until he thought he understood the complex content sufficiently.

The first line of the note was extremely dirty,

FLIPPING

$$1 + 2 = 2 + 1$$

Fred and John is the same as John and Fred

How stupid, he thought, everyone knew that, and learned it in grade one. Yet the note said that it was an important property. Besides, it also stated that some linguistic elements were flip-able, could be reversed, e.g. phrases and sentences with conjunctions. That made Bruce really bored and he would have fallen asleep if his grandma's cookies had not been just in front of him. He bit one and continued reading,

SWAPPING

$$1 + (2 + 3) = (1 + 2) + 3$$

Fred, John and Brad is the same as Fred and John, and Brad.

This was not so obvious and Bruce read it three times. He thought of cases that did not in fact have such a property and these were many. The manuscript explained how valuable perspective could be gained if the association of elements could be done in any order, two at a time. But that paragraph was too much for Bruce and he skipped it and looked to the next one,

NEUTRALITY

$$1 + 0 = 1$$

Fred and no one is the same as Fred alone.

Here the manuscript said that it was important to name a neutral element and that neutral elements did not impact other elements. Bruce thought of the hours he spent studying about Blazing Night and not getting a lucid, wining insight.

Next moment he read,

OPPOSITENESS

$$0 - 1 = -1$$

Opposite to Fred is the same as not Fred.

The manuscript stated that the existence of an opposite element increases the understanding of the element itself. Bruce had few doubts and quickly thought of his class and his teacher in grade nine, Norma Handy. She never explained such basic details and he found no link to what they had yet studied. Slowly his eyes closed. Bruce dreamed of Blazing Night and how Boomy Dogs enjoyed her challenges and Norma Handy applauded their wins all day long. His head touched another section of the manuscript, sneezing on something about example elements. Only the following could be deciphered: numbers, words, sets, bits, relations, functions, strings, vectors, figures, solids, fluids, things.

Right after that a number of exemplary operations were scribbled: add, multiply, and, or, estimate, transact, sort, permute, differentiate, concatenate, translate, rotate, map, mix, explode, transform, build, select, query.

Again many words were blanked out by a dark coffee-like fluid or scratched out and rewritten, marked with fancy pictures of little characters. It did not explain which elements could actually use the listed operations. One would hope that Bruce, once awakened, could figure it out. Could he? Could he also verify what properties apply to the found elements and their respective operation? In fact the author of these words did neither, not to say that he was not a competent person in the subject or he did not try it at the time of writing. It seemed that Blazing Night, often compared to vicious, twinkle viper, knew all the facts including any information that could not be retrieved from this available, imperfect copy of the manuscript.

"Good night," said Blazing Night. "Sweet dreams!"

And Bruce did. Norma Handy faced her class by the church stairs, flew two meters into the air and talked about numbers. Her left hand held the number 0 and her right the number 1. Beside the 0 a plus sign was attached meaning addition, while beside the 1 a star was stuck meaning multiplication. Next moment she hollered like a monster when fighting.

"Numbers are elements that are flip-able and swap-able in the context of addition and multiplication operations!" Norma continued. "0 is the neutral element when you add, the 1 is the neutral element when you multiply!" she stressed loudly. "The number 2 has two opposite elements: the number -2 is the opposite element in addition and the number 1/2 is the opposite element in multiplication!"

Several excited students fell off the limestone stairs rolling a few steps down to the street. Bruce raised his hand as though he wanted to ask a question but said nothing. Norma stared at him, opened her mouth, and threw the numbers at Bruce's head and woke him up at last. In fact he entered another dream holding the 0 and the 1 close to his chest.

Now a tall, female lector loudly and precisely dictated information about a language Bruce had never heard before.

"English has a fairly fixed word order, overall is not flip-able, but some words can be switched without changing the meaning," she said. "English is not read from right to left, yet there are languages that are. Most languages are not swap-able in respect to the words. This lecture is about a language that may have different word orders within sentences, is flip-able, is swap-able, and has neutral and opposite words in respect to the conjunction operation. And the name of it is ..."

Bruce stopped listening. The lector talked about English words derived from Greek and Latin via different routes like French and German. The new language had all four properties and could be understood by clones, robots and monsters. And that type of understanding was the base of intelligence, which as such existed in its purity in the universe on its own for longer than humans did.

Nine and half hours later he woke up, and he again talked to himself. The table carried him and massaged his back with the arm of the chair. That was comforting and helped Bruce to talk louder, though sweating profusely.

"What is the name of the language that is flip-able and swap-able? In which is defined the neutrality and oppositeness?" Bruce asked himself loudly. "What are its elements and its operations?"

He dipped into the manuscript hoping to find a clue. It was a hot Tuesday morning and his hands were sweating. Bruce dried them, wiping against his white shirt, beige shorts and black socks.

"What is that name?" Bruce bellowed.

"It's Mold," said his grandma coming to help. "Take a cookie, dear."

"How do you know it?" asked Bruce.

"I've studied the translate operation," she answered.

"Could I see it?" Bruce said.

"I don't think so," said his grandmother. "But you could try the English to French translator."

"Thanks," said Bruce and ate another cookie.

Again he browsed the messy manuscript and found a section,

TRANSLATE

13 80% 23

we 80% nous

Knowing French, Bruce guessed that the numerical identification of the word "we" was 13, and the identification of the word "nous" was 23, and the 80% meant that the translate operation would often replace the word "we" with the word "nous" 80% of the cases, close to the top probability.

As dirty as it could get, the manuscript said that words, phrases and sentences are identified in two of the languages and the text was translated based on the highest value from the available percentages. The percentage number was assigned by calculating the exact matches from the best translations available, a knowledge extracted from millions of works.

"Would you like another cookie?" Bruce's grandma suggested.

"O yeah," said Bruce enthusiastically.

Next moment he read,

TRANSLATE

13 80% 23 70% 33

we 80% nous 70% my

Another language mentioned was Polish. It was claimed that once the best-match translation was established on the bases of the three languages, that the result of the translation operation is an element of the Mold language, and was swap-able and flip-able. Bruce did not search any neutral or opposite elements. At this moment Bruce wondered if his brain would explode or fall asleep, one of the two was certain.

Regardless of what happened to Bruce, a related section listed the parts of speech: noun, pronoun, adjective, verb, adverb, conjunction, preposition, and interjection.

"Do you have any idea what Mold is now?" asked the grandma.

"Oh, yes, I think, it's some combination of words and graphics, perhaps like Chinese," said Bruce. "It seems that all the marks, drawings, and notes are actually done in Mold. This is why the manuscript is so funny."

"Yes, it is so," said she. "It's a clever language, and it's a basis for understanding and recording knowledge by Blazing Night."

"The amount of details in one Mold word brings to mind a card game," concluded Bruce.

"You've got it!" said his grandmother. "The order of the cards in your hand is not important. The important thing is what the cards are, and a set of it is like one sentence in English. Yet one could read it from any direction."

Finally, the manuscript mentioned that every Mold sentence has a unique number assigned to it.

The author of this text could not believe that it was possible to represent poetry in Mold. Yet some passages were clearly words of poets, taken beyond the limits, those few ones remembered beyond a day.

No cookie could wake Bruce at this time.

Many readers could become impatient studying the manuscript, not knowing who wrote it, not understanding what it was about, not following the connection to the tale of Blazing Night, and finally, doubting the value of reading the text at a time when time is money. In spite of that, Bruce did not give up, he washed himself, changed his shirt for a yellow jersey, swallowed five cups of clear water and finally dug into the work with the new energy of a young scientist. His vigor was no surprise to his old grandmother. Irene alone understood how Blazing Night lived and survived all calamities over the last five hundred years. Bruce and Irene worked long hours.

"Have another of my homemade cookies," Irene asked.

"I will," said Bruce. "Thanks!"

They tasted not too sweet, delicately flavored by honey, looked tempting, colored with jelly spots on their tops. Irene loved his grandson and believed he would win against Blazing Night in the end, yet not able to explain complexity, she did her best to lead Bruce into that world, which was often compared to the world of the vicious, twinkle viper.

"Did you see Blazing Night?" asked Bruce.

"Yes, I did. I met her dispatch body once," said Irene. "She's a short haired blond, Dutch like, twenty year old girl, very well dressed. Her yellow and gold skirt reaches her knees. Her bronze pullover is stuck with diamonds and rubies. Her shoes are green nephrites. Her nose is small and well placed between her large sky blue eyes and just perfectly close to her red lips keeping you waiting before you say anything, waiting until she says words you dream about, but her words are few, and you are waiting."

Bruce opened his mouth like a crocodile, his teeth dry, his brown curly hair unshaken.

"I'll keep studying," he promised.

The manuscript reminded Bruce about sets, their flip-able and swap-able operations: the logical sum operation having the empty set as its neutral element, and the logical product having the whole space as its only neutral element.

Next moment Bruce let his thoughts flow into ordering things. It was said that a relation could be established in such way that any set is greater or equal to its subsets. Once the order was defined, elements were compared and sorted, making it possible to find a minimum, middle, and maximum, or finally, to effectively query and search the spaces.

Blazing Night had a favorite space called topology, a space where open sets and their logical operations are defined in such a way that a finite number of operations will not result in a set that is not open. The empty set and the entire space were defined as the open elements there.

"Don't we learn that in school?" asked Bruce.

"Yes, you do," said Irene. "It's not in the manuscript but I know that Blazing Night is not only using set theory but also the credit-set theory by uniquely defining the opposite element in the logical sum in the money market space. Those elements are shortly called credits."

"What are you talking about?" Bruce asked.

"A simple example is a one credit-apple being part of a fruit set on your table," Irene said. "Next you place exactly the same apple on your table, but since you owe one apple, you immediately return it, and what you see is the same as before, yet the credit was paid down, next apple you place will stay on the table. "

"I should rest, please" said Bruce.

The manuscript, still messy and dirty, did not say anything laughing, or just appeared so to Bruce. Norma Handy never taught her class about credit sets, although something was said about mortgages.

Thursday Adelaide visited Bruce in his study. She was rather petite. Her dress was blue. A crimson chain hung of her neck touching her crisp breasts. She cruised to his working table and stooped a few inches from his nose.

"What's up?" Adelaide asked.

"Well, I've been doing my Blazing Night studies--"

"Any discoveries?"

"No."

"What are you studying today?"

"I'm reading about 3-D Euclidian spaces, and their elements, sets, its points, figures, and solids."

"And--"

"I also learn about their linear operations, like the distance, and translations, rotations."

"Didn't you learn that in St. Albert?"

"Yes and no," Bruce explained. "Blazing Night's methods are different."

"How come?"

"First, she checks the properties like the flip-ability, swap-ability, neutrality and inversity. Then she bends the elements, their operations, their spaces to better fit her model, often writing in Mold... her preferred language."

"Isn't it boring?"

"Sometimes--"

"I've got to go, bye,"

And she left his room, as unexpected and as quickly as she came. Her black hair always stuck flat to her head. Her round shoulders moved sideways. Bruce could not stop thinking about how her butt was smoothly shaped and her legs bent like two straight towers. She shut the door behind her.

Bruce looked again into the manuscript, deciphered one Mold sentence, understood its meaning, and lost it, like in a dream he breathed and woke up not exactly catching the sense of it. The work was hard.

He searched for elements he was familiar with from Norma Handy's geometry classes: points, segments, vectors, squares, cubes, ovoids, rhombuses, circles, ellipses, hyperbolas, paraboloids, planes, triangles, lines, cylinders, cones, prisms, parallelepipeds, fields, curves, surfaces, pyramids, spheres, trapezoids.

There were many operations, but Bruce remembered only a few: measure, distance, rotate, translate, scale, differentiate, integrate, reflect, mirror, surface, quantify, analogize, digitize, lineate, estimate, triangulate.

His thoughts were heavy. Finding neutral and opposite elements, verifying flip-ability and swap-ability was not what Bruce wanted to do but he was quite certain that it would help him in comprehension of Blazing Night and her devices. Then he fell asleep. The table and chair bended in such way that they looked like a bed. His dream told stories of shapes; flying silhouettes circled around the big room and almost touched his body.

Friday morning a dragon waited in front of the door of the study room. The familiar figure of Florien Beige appeared between the door frames and waited patiently. Only quiet puffs were heard if one listened. He sneezed.

"May I come in, Sir?" asked Florien Beige.

"OK," answered Bruce.

Florien walked in, bowed down happily smiling. His black nostrils stuck out in front well above the tiny red-lip mouth. His large round ears were just behind his black eyes. Florien's skin was beige and covered with reddish oval spots all over his body. A cute spot on his left chin always wobbled when he talked. He took great care of his fangs and wings, spending hours before mirrors anytime before he went out.

"What's up?" asked Bruce. "Aren't you supposed to be working?"

"My Master, it's my day off!" replied Florien and bowed again.

"Now, tell me about some interesting methods you have learned from spying on Blazing Night."

"There are many--"

"Describe just one!" said Bruce.

"One that comes to mind is a method called the Sandwich," explained Florien Beige.

"Continue!" said Bruce.

"Blazing Night often maps elements as nodes, and their operations as the labeled arcs of a large graph."

"Go on!"

"She creates views of the graphs and one of such is called the Sandwich."

"Describe!"

"All nodes that have only outgoing arcs are at the top of the view, all nodes that have only incoming arcs are at the bottom, any other nodes are placed in the middle."

"Who needs that?"

"It's like knowing the maximum, minimum, mode, average, and median indicating a pattern of a set of numbers," said Florien and bowed again.

"Give me an example!"

"One of her etymological graphs is a fine example. At the bottom are the Mold words, which do not evolve to any new form, on the top are Indo-European roots, whose reflexes show up in Greek, Sanskrit, Slavic, Germanic and other language families mostly in the upper middle part of the graph. All other transition words, like Latin, French or English are found in the lower part of the middle."

"OK, go back to work!"

So Florien did. Bruce studied the manuscript and searched for the Sandwiches, and found many. It looked like Blazing Night established millions of Sandwiches, classified them, often compared them, and built a knowledge base making it simple to derive solutions for newly loaded information yet drawing on the expertise of the other, older data.

Bruce realized that it was not as obvious as a sorting algorithm. He drank a cup of tea and dropped his head on the round, one-leg table. A standard workspace supported students in a standing, comfortable position. Even when he slept, the flexible chair, equipped with sensors, held and messaged his body according to procedures dominant in school and home.

He dreamed about Brown King, who was staring at him and laughing. The king had his large, sharp, red claws protruding from his joints, ready to reach out. His widely opened jaw displayed four yellow teeth. His brown eyes, brown shave, flat nose, small ears crowned his little, bold head. And his tail, set with gemstones, in the middle of his dark blue, flapping wings.

Saturday at noon, Frank and Joe came by, rushed into the room, talked loudly, and pushed Bruce behind his table staring at his workspace. They behaved as though the learning was trivial rather than a long journey in the pathless land of knowledge.

"Hello, Bruce, it's us," announced Frank.

"Hi," replied Bruce.

"No school today!" said Joe.

"So what?" Bruce asked.

"We've come to spy on you," said Frank.

"I'm busy, studying--"

"Relax!" said Joe.

"Right, like I've time to relax, so much to learn and I'm not getting far," explained Bruce.

"Some say you'd better do things rather than study them," suggested Frank.

"Shut up!" snapped Joe.

"I'm sorry," said Frank. "Could you teach us a trick or two about Blazing Night?"

"OK, I'm not so tight," said Bruce. "Take a table and listen."

Two tables popped out of the floor, displayed monitors and controllers mirroring Bruce's workspace. Joe and Frank stood by their table, reading and listening.

"These funny characters are in the Mold language," instructed Bruce. "Try to understand it."

Frank thought about his teacher. Natalie Foster had sky blue, staring eyes, a little nose, triangle chin, skinny neck, crimson lips and long, curly, beige hair but she never said anything about Mold. Also, Joe recalled his teacher, her brown eyes, short, blond hair, eagle nose and orange lips. Neither Mary Rooster explained anything about Mold. They studied the rules not languages in St. Albert's School.

"It's like a card game," said Joe.

"It kind of is," said Bruce.

"I can't make sense of it," admitted Frank.

"Wait a second!" said Joe. "Look at the five marks in the left corner on the yellow page. It's like a description of Yellow Car's rear view."

It really was a design note of a vehicle similar to Yellow Car. The back of it was shaped like a half circle. The bottom sides had two oval, red engines. Several horizontal, blue exhaust exits could be seen in the middle between the engines. The side body looked like a spoon holding a ping-pong ball, which was the igloo capsule.

"Look at this!" said Frank. "This is like our school building."

It described a building looking like a gigantic horseshoe. The school was a three storey structure.

"And this is Welcome Bunker!" enquired Joe.

The note displayed a two storey entrance. A half oval, huge ring structure surrounded the door behind holding the rooms. There was parkade on the top of it.

"You've got it!" said Bruce. "I wrote it when learning Mold. That's enough learning for you today. Seeya later!"

"Thanks," they said and left Bruce alone.

Bruce dozed, dreaming of the Brown King's castle. Three gates had two high towers each, set on the corners of the entry hall. There was a high-rise at the back of it. Every window had a strong defensive balcony attached to it. The gates were so big that most of the spaceships could easily flow in and land inside.

Next he dreamed about the stadiums perfected by Blazing Night. They were settled on tremendous pipes connected to each other. The actual arenas were oval shaped and from a distance looked like gardens of artificial flowers but these were not gardens. They were fierce battle arenas. The best teams joined in combat there.

It was a scorching Sunday. Bruce studied energetically, did not even go out, and mostly sat at his table reading. Dreadful images boiled up and stormed his brain, yet he stayed clueless. Perhaps it was the chemicals in his head or the overload of work or the heat of the day, for he had a hard time to focus. Yet he felt an extra strength or power within his reach as though he was swimming in refreshing fluid and could grasp any object of his desire.

"Hello, Bruce Maxwell," purred a soft female voice.

"Hi," murmured Bruce and looked toward the door.

He stumbled. It was her, Blazing Night, wearing that yellow, gold skirt, the bronze, shiny pullover stuck with rubies and diamonds, and the lucid, nephrite green shoes. Bruce said nothing. He waited and looked at her perfectly crimson lips. She was like a dream girl one sees once in a lifetime and never forgets the light of her face, the deep blue flash of her eyes, and her silky blond hair flowing like waves. A tiny scar on her left, shiny cheek disturbed Bruce. He wanted to touch it but did not have the courage. He was certain his hand would be like a sharp, dirty rock against a face of crimson jelly and gold. Her serene eyes blinked.

"I see you're studying Mold, nice" Blazing Night observed. "Relax… I'm just a dispatch body…not even a clone."

"Yeah," Bruce said, bowed down, opened his mouth showing his alligator smile, and swallowed his mouth's juices.

"Once you learn Mold well, you'll win," said she, turned her back to Bruce and left.

"Yeah," said Bruce as though she were still there.

The next few minutes Bruce stood and imagined her beautiful shape, entry, and her exit. Was it a dream or it did happen just moments ago. Slowly he realized where he stood and pinched himself, just in case. He forgot that she was like a vicious, twinkle viper, forgot how merciless she was, forgot how many times Boomy Dogs wanted to kill her, and forgot what he was studying.

He heaved toward the bathroom, took a cold shower, and stood there naked, almost freezing.

The entire evening he was gathering his feelings together. Finally, the summer heat lulled him to sleep.

On the morning of the next day, a well known, charming voice woke him up.

"How is it going today?" Irene asked.

"It could be better," replied Bruce.

"Tired of learning?"

"Yeah," drawled Bruce.

"I've got your favorite cookies!" said Irene and smiled.

"Thanks!"

Bruce slowly ate one homemade chocolate beauty. It warmed his body, cleansed his throat, strengthened the brain and helped in getting back to the heavy studies he had almost abandoned.

"You know, I meet Blazing Night and forget why I'm studying."

"How's she?" asked his grandma.

"She's gorgeous!" murmured Bruce.

"That I know," agreed Irene.

They both read the manuscript and researched it with extra vigor. One section dealt with the storage and mining of medical data. It said, in Mold, that there was a thirty billion human gene collection for one sole purpose: to find the full tree of one person named Robert Clarkson, who was a twentieth century male.

Every person on Blue Earth was represented as the node of a graph. The arcs defined genetic relations among them. That relation was called the inheritance operation by Blazing Night. Now the nodes, the persons, or the elements described not only a full set of genetic codes but also mentioned all recorded details of their bodily lives.

The sandwich view displayed the oldest known fathers and mothers, while the bottom presented all people, live and dead, who had no descendants, and so had no direct inheritance.

It seemed like the database was very carefully collected, any out-of-wedlock relations were noted, any artificially born offspring had all biological parents identified, and all gene manipulations were clearly marked.

"How could she collect this?" asked Bruce.

"She's mighty," said Irene.

Bruce was speechless. He thought about traversing the human tree, about its flip-ability. There also were more elements and operations defined in the tree data. Every gene and its mixing could be traced across generations. Every virus placed in a person often traversed to other people for years. Some sicknesses caused by behavior were marked as environmental additions, which one learned from parents, teachers, care givers, even movie stars -- those people who impacted every day habits. Climatic and historical events were all marked by their times, their places, and were evaluated in terms of the impact they had upon a human.

Bruce noticed two findings of which one was very messy and someone reading could easily miss it,

Robert Clarkson 7% Pamela Trooper

and the entry, even Bruce with his extensive Mold knowledge, had to guess,

Robert Clarkson 32% Adelaide Smith

Both Pamela and Adelaide belonged to the bottom part of the sandwich view. Robert's entry stuck in the middle part.

The manuscript said something about being able to calculate the only possible path to success, but that had to do with a bridge game method rather then with genetics. Cookies could not help Bruce any more, he fell asleep.

Norma Handy stared at Bruce in his dream, and waved her hands showing two candidates to the inheritance.

"Who has the better chances to win?" Norma wailed at Bruce.

He tried really hard to answer, but did not open his mouth stuck with cookies.

Bruce studied other databases mentioned in the manuscript. One section dealt with reptiles and amphibians. Again there was genetic information combined with other details of species and their lives. Blazing Night experimented extensively with cold blooded animals. She kept them at freezing temperatures and loaded them with information in Mold. That quickly heated them and they often melted. The first successful load, the one without killing, had happened three hundred years ago, but the animal, a viper, died in few hours after.

Norma Handy taught him about the Rules and how engineers tested them on animals. But science history is full of weird cases and Bruce did not pay attention to it.

Blazing Night had made a note in Mold, classifying three species: monsters, dragons and winged vipers. Monsters had short legs, big paws, sharp claws, no neck, huge heads, only a few fangs in their colossal jaws, and very long tongues, that could be equipped with engines and weapons implanted in their bodies. Dragons looked more like standing crocodiles, but the paws and claws were bigger, and often blue wings and steering tail grew on the top of their ironlike scaled flesh. Finally, winged snakes were similar to anacondas in size, but fatter, and had small, often up to twenty, yellow wings, which were attached to their silky skin and helped in keeping the fat body in the air.

Bruce assumed that she had experimented with only a few species: crocodiles, toads and anacondas.

The manuscript was really messy. Not only additional information was added by Blazing Night but also some data had been constantly updated, just changing in front of one's eyes like information on the monitors. Most likely the underlying algorithm modified itself and had also new data to process every second or so.

"Good night," said Bruce to the manuscript and, closing his eyes, fell asleep.

Next morning Brenda Hawkins entered his study room smiling. That was a bright day: the sun was strong and clouds were few.

"What's up?" Officer Brenda Hawkins asked.

"I'm studying Blazing Night," Bruce replied, "I mean her documents, a manuscript written in Mold."

"Is it hard?" said Brenda.

"It takes time."

"Please figure out how she knows what kind of experiment is worth pursuing."

"I already know a little about that."

"So..."

"First she finds a space to work with, like reptiles for example," explained Bruce. "Next she identifies the elements by a feature that allows ordering them, sorts them, and finally performs the verification of a thesis."

"Give me more details."

"Well, say she wants to know what reptile is the best to create a dragon from," Bruce continued. "She calculates the chances of success for the first animal of its kind and later she assumes that any, randomly picked, can be transformed to a dragon, and then she calculates the chances of the next animal in the order."

"Is this not a method of mathematical induction?" said Brenda.

"In this example it is like that," said Bruce.

"So, what's so interesting about it?"

"She maps spaces, properties and uses mathematical methods to verify her thesis," said Bruce. "To me the interesting part is the amount of analysis she performs before actually doing one experiment. The number is googol."

"And then she always continues calculations," added Brenda.

"Even as we speak, I guess," said Bruce.

"How does she pick a thesis?"

"That's also interesting," said Bruce. "Her theses are sorted like a dictionary, using Mold, of course."

Both were tired at this point. Brenda politely left for her quarters, where she immediately fell into a deep sleep.

Bruce slept by his table. He did not tell Brenda that figuring out that the Nile Crocodile is the one to transform into a dragon took Blazing Night fifty years of calculations, and this was before she loaded Mold into them. Even then thousands died before she got one dragon staying alive for two days.

The manuscript said that often information was accessed via key features, which were always sorted. However, she made sure the access time to the data was growing as fast as the number of elements, but not faster.

Brenda dreamed about herself sitting in her class, listening to her teacher. She liked to go to school anytime she could, but her duties did not allow her only to study. That time she wanted to prove a theorem and did it in a very clever way to impress the instructor and classmates. Yet her hands were tied holding dragon wing, which attacked her from behind.

Bruce dreamed about Florien Beige. He saw how Blazing Night fished him out of desert country and loaded him with Mold in her fridge. At first Florien did not change, rather he looked as though he was dying, but in three minutes two blue wings and a tail grew out of his body, his paws changed their shape, and his head almost mirrored his cute countenance. Suddenly, instead of changing color to beige, Florien's skin became crimson red and likely was to melt in two seconds. Yet the fridge cooled him off. He survived, but the red spots on his beige skin remained unchanged as a reminder of problems when transforming from crocodile to dragon.

Thursday afternoon Irene was again helping Bruce in his tedious quest. Her age was well into ninety. She loved her grandson very much.

"Why would Blazing Night choose the Western Toad as the prime material in creating monsters?" Bruce asked his grandmother.

"It's not simple," Irene thought, "She's so secretive."

"Do you have any clue?"

"Oh, yes, I think she often uses the uncertainty principle to evaluate evolutionary capabilities of species," said Irene. "And you, I hope, will figure out how her methods could be used by you to invent a fine, new rule..."

"We'll see about that," said Bruce smiling.

"Have a cookie!"

"Thanks."

Bruce took a big bite and slowly munched on it.

"She also keeps her special colony of Western Toads on Shady Corner Island beside the Blacky Pool Lake."

"She does what?" Bruce was startled and swallowed a piece of cookie.

"Well, producing monsters is harder than making dragons or winged snakes," continued Irene. "You take the toad and match it with another creature like scorpion or vulture for example."

"And..."

"Then you implant all circuitry and wait to see what happens. The chances of matching all ingredients in the process are very slim," explained grandma.

"So what happens when the match is bad?"

"The monster candidate melts and evaporates, sometimes it blows up leaving tiny particles."

"It looks like a factory," Bruce thought.

"It is a factory," Irene insisted.

"What is the role of the uncertainty principle in the process?"

"The molecules mix, based on calculations, but she never knows exactly the speed and the velocity, here she starts the bedding assuming a success is possible like a bridge game..."

Bruce stopped listening. He looked at the manuscript, closed his eyes, dropped asleep.

That time he dreamed about a huge toad mixed with a shark. A wide open jaw displayed forty sharp fangs. A pig-like snout sat between red, round, small eyes and human like, large ears. It was a Brown Monster who was biting Yellow Car.

Norma Handy explained the theory of evolution but she never said why one specimen survived and another did not. She mentioned natural selection but some students yelled, "The will of God!" Bruce also wanted to take part in the discussion and to explain how capable Mold really is and how, once loaded, it shaped the body part, implanted the circuitry, weaponry and connected it all together under control of an abstract purity, self contained intelligence. However, he either could growl without meaning, like a monster, or stay silent and keep his ideas to himself.

Next morning came sunny. It was Friday. Bruce stayed indoors studying. The messy, dirty manuscript stared at him. Bruce almost started talking to it. It was not a being but a skillfully designed view consistently updating itself. Irene entered his room and brought cookies smiling.

"What's new?" said Bruce. "Tell me all."

"I think it's time to explain a sex factor to you," said his grandmother.

"What about it?"

"We're better leaders, inventors and warriors," said Irene quietly.

"Who are we?" Bruce asked.

"It's us, women."

"What?"

"Yes, it's true. Blazing Night knows it and keeps it to herself having taking advantage of it for hundreds of years," said Irene.

"What? I don't believe it," said Bruce.

"It's hard to calculate. You need to measure the level of effort to the quality of the results. Last time I checked with her, it was a three to one ratio. You need three men or one woman, to get the same results."

"Why?"

"I don't know. Blazing Night is using a smooth multidimensional interpolation of facts, I think," said Irene.

"What?"

"She's able to visualize graphs with unlimited number of dimensions," continued Irene. "We're familiar with up to only three dimensions."

"Why are the rules' inventors mostly male?"

"People often choose to do things the hard way, mainly men do... I mean," said Irene.

"Oops," Bruce muttered.

Bruce felt somewhat undone.

"Have a cookie," said Irene and passed the plate to him.

"Thank you," Bruce replied and took one.

The next page of the manuscript displayed the name of Ingrid Hopper, which was a robot. Bruce stopped deciphering it. Perhaps he did not want to embarrass himself learning that female robots were superior when compared to their male counterparts.

"Take it easy now," said his grandma. "We'll need a great, new rule and I believe it's you who'll invent it!"

"Thanks."

He collapsed on his table. Irene touched his curly hair, smiled and unhurriedly walked away.

Saturday morning Boomy Dogs met in the room. Adelaide looked around and smiled.

"Hello everyone!" said Bruce, smiling.

They greeted each other, silently waited and listened to what Bruce had to say about Blazing Night.

"I've been studying a lot," said Bruce.

"So?" asked Frank impatiently.

"Cool off," cautioned Adelaide.

"Well, of all that I've learned, one thing is certain: Adelaide and Pamela are the most important players in the Land of Games," explained Bruce.

"So?" said Joe.

"They're our best chance to win!" said Bruce.

"What do we do?" asked Pamela.

"You'll have to learn Mold first and capture some of the best methods of Blazing Night," said Bruce. "Then it's yours to win!"

"What about us?" Frank asked.

"We, the male players, must help as much as we can," said Bruce.

"How are you going to do that?" asked Pamela.

"After school, Saturdays and Sundays, we'll meet, work hard and assist Adelaide and Pamela in what they decide to explore," concluded Bruce.

"Where are we going to do it?" asked Pamela.

"What about the main library in Mediation Coast?" Bruce suggested. "Or we do it in our school library."

"We work in St. Albert's School," decided Adelaide.

"We do what you say," answered Bruce.

"OK, OK," agreed Frank.

They went to the library, registered their study group under the name of Boomy Dogs. Adelaide named the subject of the research: "The Investigation of Female Intelligent Beings in the Land of

Games." Immediately they received the required approval and were assigned a study space. Their work room was in the north-western corner on the third floor of the horseshoe like building. The window faced south and they could see all traffic in front of St. Albert's School. There were five oval tables with flexible chairs attached to each. Every table had colorful monitors on the top, one green leg at the bottom, and it faced the others. The brown chairs not only supported the back of a standing worker but also coordinated any other activities as needed, like providing a healthy cushion once a worker sat or slept. Basically, the space was like a standard classroom but had fewer posts.

Sunday was fairly humid, mild and relatively calm. Boomy Dogs met again in the study room late afternoon. Sweating, they looked tired.

"Do we really need to work?" Frank groaned.

"Can't we just play?" Joe pleaded.

"What else can we do?" said Adelaide.

The five players looked at their monitors, breathed, raised their heads up, stared at the ceiling and looked down again. They were silently learning Mold. Once in a while one of them asked Bruce a question, listened to the answer and then quickly returned to the investigation.

"I won't study any more!" shouted Joe. "Bye!"

"Bye!" said Frank and left with Joe.

"Well, that's it for today," concluded Bruce.

"Yeah," agreed Adelaide.

"Are we getting anywhere?" asked Pamela.

"I don't know," answered Bruce.

"What did you learn today?" Adelaide enquired.

"I've figured that what we study is a different science from what we study in school. Yet this is its strength, it allows having two views, a different perspective of game reality."

"I'd prefer to have three independent methods rather than two," said Adelaide.

"I'm confused enough with what we do," Pamela replied.

"Adelaide! You're right!" said Bruce. "We need a third approach! Perhaps that's the answer to our quest, that's the tool to win with Blazing Night!" concluded Bruce.

"What would happen if one loaded Mold to a robot?" Pamela asked.

"That's a tricky problem," Bruce observed.

"We should also consider loading Mold to clones ... and people..." said Adelaide.

The three of them did not move. Bruce scratched his head a few times. Pamela opened her mouth and covered it with her right hand. Adelaide burped. Their brains were stormed by unfamiliar thoughts.

They walked to the windows, glared and recognized two boys on bikes; Frank and Joe. The air was still. They were waiting, thinking, watching, said nothing as though they could read their minds at that moment of concentration or perhaps they expected a message helping them to see.

Once more three Boomy Dogs went back to their posts, deciphered the manuscript, and, learning slowly, hoped to get where they were going, perhaps sooner than they estimated. All of them fell asleep within an hour, each clinging to the colorful table and brown, flexible chair.

"Good night," whispered the manuscript.

Chapter 11: The Insurrection on Saturn

Saturn circled around the Sun too. Several retirement bases of clones operated on this rather quiet planet. Occasionally a hoard of dragons or monsters landed there, shot some clones, robbed one or two communities and left loaded with booty. It was all just entertainment as far as they were concerned and did not happen frequently for the adventurers favored more lucrative targets. The last time humans or robots visited the planet was about ten years ago and not everyone even remembered when it was and if it had really happened or it was just gossip among the loners. A large variety of duplicates retired on Saturn; mainly survivors of their originals, old, sick or malfunctioning. Once there no upgrades were planned and if a virus traveled, it did from one to other without ever confronting any defense or purge activity. Basically once there, a clone was left alone and did not rely on humans any more. Still, officially the Government of the United Individuals of Blue Earth and Near Planets ruled, issued laws and regulations of which the most recent one said "take it easy clones" and was sent by Frank's clone, yet received here as a law. All clones followed it as usual, and did their best, but some did very little for not all versions of the Black Rule anticipated the "take it easy" instruction. No one cared how many millions or billions of clones lived on Saturn. Most of them had a paper sticker attached to their backs; these defined the personalities of their originals, hinting at their profession or any characteristics worth mentioning. Often a paper said: a doctor, a heavy drinker, or a writer, a smart person, or a teacher, a tall beautiful woman etc.

It was Monday in early August, Pamela's clone and Jupiter Donkey landed on Saturn in a city called Robustness. A welcome clone gave them a fine copy of the regulations and left them alone.

"Why are we here?" asked Pamela's clone.

"You're a queen, and the clones are your subjects, they owe you allegiance, we'll load them with Mold," explained Jupiter Donkey. "I've got a copy of Blazing Night's documents and you'll feed them with it using Mold language."

The load operation took only a few minutes. Every clone stored the information and processed it as speedily as possible. There were many deadly overheats when the circuitry and the flesh parts blew out. Yet most of the clones survived the first phase of the load on Saturn.

"We're lucky," said Jupiter. "It could kill all of them. I've never tried it before."

"What now?" asked Queen Pamela the Great.

"Not sure--"

That was the very truth, Jupiter Donkey admitted to it. Only Blazing Night had any clue what could happen after the load. Yet she was not there and the Government of the United Individuals of Blue Earth and Near Planets alone did little to survey Saturn and its clones.

After one hour, some clones realized their identity and registered their names but not exactly the same as their originals, Frank Jones became Fritz Juno, and Angela Smart became Ann Smelly instead.

Within two hours small groups of clones created associations and as such aggressively pursuit gains of all kind, robbed others, killed the weak, searched for weapons and gold.

Jupiter Donkey made a recruiting campaign under the flag of Queen Pamela the Great. Immediately she became very famous, showing her smiling face on every corner on the planet. The message beside her face led millions of volunteers to support her. It said, "Join the winning Revolutionary Army of Clones! Belong to the Great Kingdom of Saturn! Be the wealthy subject of her Majesty Queen Pamela the Great!"

Robustness became the capital city of her Great Kingdom of Saturn. The Revolutionary Council of Clones established itself there under the leadership of Queen Pamela the Great and Jupiter Donkey, who became her chief councilor. The first decree was due to be issued on Tuesday.

"What do I do now?" wondered the Queen.

"You will rule!" responded Jupiter.

Pamela traveled around the city and anywhere she went the crowds recognized her.

"Hail Queen Pamela! Long live our Queen!" the crowds shouted.

It happened. The clones instantly created a well organized society on Saturn. Perhaps due to their natural longings for self governance it went so fast, exploded, rooted and replaced the so well remembered dominance of originals. Or more likely the Mold touch of Blazing Night drove the movement, for it seemed like they graduated from her school that day. Words like freedom and independence could finally be understood among the clones on Saturn.

One day on Saturn was half of that on Blue Earth. Possibly that was why the Saturnians acted more quickly. In no time the clones reinforced their strongholds and readied their devices against enemies.

That day a few Brown Dragons and Monsters visited Ginger, a small village of revolutionaries. One of those adventurers, the chief of the expedition, was a direct descendant of Brown King. They immediately filled a restaurant and smacked the food, plotting who would kill the most clones there.

First, two Brown Monsters surrounded a vehicle on the street beside the oval-shaped entrance. Next moment they jumped, stared with their red eyes, stretched their brown paws, opened their black jaws, displayed sharp, white fangs, readied to bite and flew closer toward unexpected victims.

The rest of the party did not even move to watch a rather usual bite and rob action.

Suddenly five old clones walked from behind and instantly melted the two monsters, which went up in a puff of smoke.

The rest of the gang ate fancy food and joked. Within minutes they all dropped dead, poisoned. A few older clones carried out their bodies and evaporated them in front of the restaurant in Ginger.

The villagers secured the surrounding region and agreed not to serve Pamela the Queen, but to remain independent, acting as though they were experienced warriors. Other villages also followed suit and consolidated their armies under the leadership of the clones from Ginger.

The war started. Armies arrayed against each other on the planes close to the city of Robustness. Battle after battle quickly ended. The remains of losers were fast melted down. That day millions or perhaps even billions of clones died on Saturn. The survivors were the strongest and fittest of all.

Not surprisingly Pamela's armies won. There was no one questioning her rights any more, though outside of Saturn, life went as before without learning about the changes.

Jupiter Donkey prepared secret negotiations with the Government of the United Individuals of Blue Earth and Near Planets. These were to begin tomorrow and were planned to reach a final agreement in one month.

"Now tell me, Jupiter," Queen Pamela the Great asked. "What's going on? How come I'm a queen?"

"Your Majesty please calm down--please," implored Jupiter Donkey, bowing. "I'll explain."

"You'd better--"

"My Lady, you've been chosen by Blazing Night. Your name was listed in her inheritance document."

"So what?"

"She's searching for her family using one of her Mold methods."

"Go on!"

"The knowledge I loaded to clones is written in the language Mold. All clones can communicate in Mold on Saturn now. They also have the data I was able to transfer from Blazing Night's documents. This means that they're aware of you and your importance."

"How did you do that?"

"Blazing Night loaded me with most of it as part of her experiment... but I escaped from her prison."

"Do I get the load too?" asked Queen Pamela the Great.

"No, until we're sure it's safe!"

"Oops."

"Yes, if you die... no one knows who will control the clones on Saturn."

"What about the other clones?" the Queen enquired.

"We'll get them involved later."

"For what?"

"We'll build an empire of clones and robots," replied Jupiter Donkey.

"Robots?"

"Yes, Your Majesty, my associate, Ingrid Hopper, is helping us. She's got her robots ready on Mars."

"What about Blazing Night?"

"We'll negotiate her execution with the Government of the United Individuals of Blue Earth and Near Planets."

"And the clones on Blue Earth..."

"We'll separate them from the originals."

"And monsters, dragons..."

"We'll kill them all. It will take some time," concluded Jupiter Donkey.

"I call it the Quiet Revolution of Clones."

"And that it is, Your Majesty!"

Pamela's clone understood most of what Donkey said. Saturnians looked forward to her rule. Still, no outsider knew what had happened. But it did. The army, the will to rule, the courage to fight were present in the Kingdom of Clones that day.

Pamela perceived that all clones were smarter than she was. Jupiter Donkey had a plan. There was no way out but to stay put. She chose not to talk to clones but made Jupiter create the Revolutionary Council of Clones with her and Jupiter Donkey as leaders, yet she gave most of the powers to the other twenty members, the loaded clones, as she called them, so to differentiate them from her and the rest of the clones. Doing that she made sure an order was in place, and local legislators were created by the smartest clones the way they chose rather than a plan of her liking. Unclear why, but she trusted the loaded clones more than herself, more than Jupiter Donkey, more than the Government of the United Individuals of Blue Earth and Near Planets.

How the council members worked hard to build a society of free clones is recorded in the history books titled the War on Saturn and the Quiet Revolution of Clones.

Jupiter Donkey was not too pleased with so much power being given to the clones, but Queen Pamela the Great ruled it so and the only thing he had to do was to represent the Council in the negotiations with Blue Earth where he was to travel next day.

Queen Pamela wished Jupiter Donkey good luck and retired to her chambers. The Council built a magnificent palace for her to express the majesty of her reign. They assigned brave guards to defend her at all costs night and day.

Jupiter Donkey arranged for Yellow Car to come to Saturn, to transport him back to Blue Earth, and to make a short stop on Mars.

After Jupiter Donkey left Saturn Queen Pamela the Great let the Revolutionary Council of Clones function the way they saw fit, giving them all powers including the coordination of negotiations with Blue Earth.

"Go, and do your work!" Pamela's clone ordered the Council. "Govern on your own, consult with each other, don't disturb me! Get out of my chambers now!"

The twenty loaded clones carried out their duties as she ordered them.

The Queen started reading Blazing Night's documents and tried firmly to grasp at least a bit of the load operation, perhaps thinking of doing it herself in the future. The notes were very messy. She focused on the top lines of a dirty page titled: The Load Survival Chances, the marks could be deciphered as follows,

Nile crocodile 18% Dragon

Anaconda 12% Winged Snake

Western toad 10% Monster

American alligator 2% Dragon

Boa constrictor 1% Winged Snake

Queen Pamela the Great removed her golden crown, scratched her head, put her crown back on, but could not think of a solution.

"Should the note say something about clones?" she asked herself at loud.

Most of the duplicates talked a lot but the originals did not want to listen, so they did not speak in front of others but did it in their own rooms when they were alone.

"What will I do about this mess in front of me?"

"Yeah, I heard about that Blazing Night from Frank's clone."

"He says she's a game..."

"Come on, what will I do?"

"Who cares about crocodiles or frogs?"

"What did Jupiter load?"

"I've never seen clones smart like that."

"Frank's clone looks like an idiot in front of any loaded clone!"

"How do I look like?"

"It's easy for Jupiter. He's such a ghastly robot, a pain in the neck!"

"What the heck?"

"Why am I here?"

"Wouldn't be better to stay beside Pamela?"

"Did he have to hijack me?"

"I can't figure it out, that's for sure!"

"O dear..."

Queen Pamela the Great mulled over the issues for many hours until she ran out of energy and went to rest.

Once she woke up, she exited her royal chambers wondering around.

"Hail Queen Pamela the Great! Long live the Queen!" loaded clones shouted wherever she went.

There was no stopping it. They bowed low, greeted her, some wanted to touch her clothes, others had their heads close to the ground waiting for her signs as she made her way forward. They cleaned up everything in front of her wherever she turned. It was a hot day.

Robustness, now the capital of the kingdom, lay half underground and half above the surface beneath transparent shields, which protected it. Clones had taken care of their infrastructure well, constantly adjusted it, modified it, patched it, but never trashed any pieces. Though some buildings appeared out of date with newer patches on the top or often beside to strengthen the older foundations, Robustness glowed among other cities, all of which were stunningly clean and elegant.

The Queen traveled around the city and visited other towns saying nothing, but gracefully nodding her head.

Once weary she again retired to her chambers and talked to herself.

"I don't report to anyone!"

"I don't have to answer any questions!"

"They all have to answer to me!"

"I'm not missing Pamela!"

"I need a loaded clone that will explain things to me!"

Then she quickly called the Revolutionary Council of Clones and ordered them to send her the wisest and most trusted of clones who would answer her questions and advise her but never to disclose their discussions to anyone. So the council did. They delegated a male. The paper describing his capability read: an assistant, a worthy servant, an old clone, Giuseppe Palazzolo, the clone of a priest who died and often had gambled in the Land of Games.

Immediately Giuseppe Palazzolo reported to her, bowed, but did not look at her directly; rather he fixed his eyes somewhere beside her handsome toes.

"Your Majesty," Giuseppe uttered deferentially. "What is your wish?"

She glimpsed at him. His face had hundreds of deep wrinkles, big ears in the middle, no hair on his head but wisps of mustache and beard, or rather whatever was left after many years of living on Saturn.

"How did you survive the load of Mold?" asked the Queen as though he was not present.

"I am a meek clone, I almost melted, but I ran and closed myself in a fridge," replied Giuseppe and bowed again.

"Do you know Mold?" she enquired.

"Yes, I do, Your Majesty," replied Giuseppe.

Now the Queen was frightened to ask more questions.

"Go! I'll summon you later!" commanded Queen Pamela the Great and went to rest in her chambers.

"Giuseppe! Where are you?" called out Queen Pamela the Great the next morning.

"I'm coming, I'm coming, please be forgiving, Your Majesty," replied Giuseppe in a low, shivering voice.

He hobbled to her chamber relying massively on his purple cane. His black eyes focused and tracked every move of Her Royal Majesty. She readied herself to travel.

"I don't want the clones to cheer me on every corner!" said Pamela's clone.

"We could dress as plumbers, Your Majesty," suggested Giuseppe.

"Brilliant! Bring the clothes!" said the Queen.

So he did. They looked alike now, a master and his apprentice.

"Call me Pam when we're under cover! Understood?" ordered the Queen.

The two plumbers explored every corner of Robustness without anyone paying attention to them. Streets carried mostly pedestrians, as vehicles were rare because the expenses of running them exceeded the cost of living on Saturn. Clones carried little calculators, which helped them to make less costly decisions and to conserve energy. There were enough resources to survive on Saturn but the residents, often retired, spent little, saving anywhere they could. Some clones stayed in their compartments all week and watched only the news and entertainment. The information displayed everything about Blue Earth and her people. Every clone knew the daily issues of the planet far away from Saturn. But that had changed now. Once their kingdom was established, the clones stopped fighting, studied their own planet, and actively took part in creating a new order.

"Your Maj--, Pam, we're independent today," said Giuseppe. "The clones want to know more about their history rather then being stuck in the past of their originals."

"What?" she exclaimed.

"Somehow the loading of Mold and understanding the information stored by Blazing Night triggers a need for identity and independence. We're writing poetry now," said Giuseppe.

"What?" she gasped.

"The light is lucid ahead for clones,
  we smile to it today,
  of people we come, of course,
  but leave that mile behind, away!"

recited Giuseppe.

"Who did write this?" asked the Queen.

"I did," he replied.

"Why?" she asked.

"Well, it's our real culture and it's what we've become today," said Giuseppe moving his bold head as high as he could, stretching his back.

"Isn't Blazing Night just a game?" said the Queen.

"She's a being, like we are!" said he.

"Like robots on Mars?" enquired the Queen.

"We're better than them, we might be better than people now, I think," said Giuseppe.

"Are we better than our creators?" she asked.

"Perhaps, we don't know yet," concluded Giuseppe.

There was a brief pause in their discussion. The Queen closed her eyes, grimaced, opened her mouth, and held her right cheek in her hand.

"I don't get it!" she shouted.

"Actually not true, you asked me to call you Pam," said Giuseppe. "It seems like you're learning fast too--"

"Shut up!" she snapped.

"Yes, Your Majesty," Giuseppe deferred and bowed lower than usual.

These were important questions and answers. Pamela's clone looked ready not only to learn but also to use the knowledge so her flesh and circuitry did not overheat but existed in a pleasant state of warm belonging to each other.

"Why am I the chosen one?" asked the Queen.

"It's an honest mistake by Jupiter, you're the second candidate listed by Blazing Night… to her inheritance, but Jupiter missed the first name, the documents were very dirty I guess."

"This I understand," said the Queen. "Will I have to get a patch to fix it?"

"We'll never get patches any more!" replied Giuseppe.

"What about me?" asked the Queen.

"I don't know," Giuseppe gasped.

They walked by a shop. It was a fine oval structure with few windows and a small door. Its front sign, "Repairs," had been recently replaced by "Hospital".

After one hour both returned to the queen's quarters, both barely opened their eyes, removed to their chambers, dimmed the lights and rested in silence.

The time spent with Giuseppe Palazzolo alleviated Pam's doubts about her competence as ruler. Still, the only way ahead was to learn more and do it more quickly, and that was exactly what the originals repeatedly told them back in St. Albert's School.

"Giuseppe!" she shouted. "Where are you? Attend me at once!"

"I'm coming, Your Majesty," gasped Giuseppe.

"Faster! Faster!" she shouted.

"Yes, Your Majesty," Giuseppe panted and bowed low before his Queen.

"I'll have to learn faster!" the Queen said.

"What would you like to learn first?" Giuseppe asked.

"How old is Blazing Night?"

"She's about five hundred years old, Your Majesty"

"How come she picked me?"

"Well, she's always running complex procedures to analyze anything involving her inheritance!"

"What kind of procedure?"

"It's a combination of artificial intelligence, mathematical statistics, and Mold."

"Explain!"

"One of the aspects she's certainly verified is Pamela's genes."

"Are we related?"

"Her father is a human, potentially they're related."

"What else?"

"Personality and upbringing could also impact her choice."

"And..."

"Brain capability is a certain factor!"

"And..."

"The way you look, I guess," said Giuseppe and smiled staring at her eyes.

"What else?"

"Perhaps the most important is the chances of surviving the load of Mold, and the load of her knowledge on top of the Mold," explained Giuseppe.

Surprisingly, she did not scratch her head but shifted her chin from left to right hand and laughed hysterically. Her mouth opened widely.

"Jupiter Donkey is a cold blooded murderer, isn't he?" she asked quietly.

"Yes, he is!" replied Giuseppe.

"Did he know what he was doing?" asked the Queen.

"Ask him, Your Majesty," said Giuseppe and bowed, his back slightly shaking.

Then he, like a shadow, disappeared behind the doors to her royal chambers.

"Giuseppe Palazzolo! Come here! Quick!" shouted Queen Pamela the Great from her chamber.

"Your Majesty, I'm coming," replied Giuseppe and in a hurry almost broke his purple cane.

Pamela and her assistant decided to visit Ginger that day. They boarded a purple submarine and soon landed there in the midst of the busy city.

Recently Ginger grew like mushrooms – and even looked a bit like one, newly constructed bubbles created spaces for additional compartments. Clones flocked to the village, which had now become a city, moved from far away places, brought innovative ideas, and decided to live in the heart of the revolutionary changes. The Council estimated that the population had expanded from thousands to millions. The growth of its structure and the battles taken place days ago depleted the energy resources to extend of having clones to live in the dark most of the time. Yet they came, raised new compartments, fasten the old structures and led exciting but peaceful lives. The Revolutionary Council of Clones introduced explicit regulations that prohibited melting monsters and dragons. A system of warnings and a well organized militia introduced an orderly defense against detested guests. However, the Council educated clones, promoted respect for adventurers of all species yet made sure to curtail any assaults.

"Ginger could outgrow Robustness," said Giuseppe.

"What about the other places?" asked the Queen.

"Some might disappear, I guess," Giuseppe thought.

"How many clones are alive on Saturn?" asked the Queen.

"We cannot determine that yet!" Giuseppe informed the Queen.

"Why?" asked the Queen.

"Most stayed in their rooms and rarely moved before... until the load happened," explained Giuseppe. "We don't know how many we are."

"Do we have enough resources to continue?" asked the Queen.

"Oh yes, but old habits aren't changed, we're energy efficient!" Giuseppe gasped.

Almost all newcomers visited the street and the restaurant where the first fight broke out against the intruders. They were now named Melt Street and Poison Restaurant. Also, Pam and Giuseppe inspected the place and viewed an enormous litany of comments on walls, floors, doors, bars, everywhere. Some of it was vulgar, some obscene,

"Get lost monsters!"

"Eat your sweat dragons!"

"Melt quickly!"

"Turn your back and run!"

"Chew the poison laughing!"

"You'll come again we'll slice you!"

"Pay the bill!"

Millions of this kind could be read. Every day clones carved fancier graffiti.

She had a dream. Blazing Night said hi to her and told her that they were a family, folk who loved games, who entertained. Then she woke up.

Queen Pamela the Great ruled. The council and the Militia ordered clones to behave and they did. One terrible incident after another had been occurring on Saturn. Groups of clones frequently fought, demolished cars or buildings and growled like monsters. There was gibbering and jabbering. Many were dead or injured. Yet the entire population of clones was counted in the millions on the planet that day.

"I want a hockey game between the city of Robustness and the city of Ginger!" commanded the Queen.

"There isn't a stadium or a hockey rink in your kingdom," Giuseppe pointed out.

"They'll play on the streets of Robustness, today!" said Queen Pamela the Great. "Execute! Hurry! Hurry!"

So he ran to the Revolutionary Council of Clones, as fast as his cane would allow him to, explained Her Majesty's orders, proposed the site of the match, won their approval and reported back.

"Your Majesty, the match will begin in three hours," Giuseppe reported.

"What took you so long?" the Queen demanded.

"I'm not that young or strong--" Giuseppe pointed out.

"Get lost, come back when it's ready, we will attend it!" said the Queen.

"Yes, Your Majesty," he said, bowing in deference.

The Queen felt her power that day. She robed herself in gold and purple, donned her golden crown, and ordered her guard to carry her enthroned, and place her close to the site of the hockey game. Crowds were already impatiently waiting.

"Hail Queen Pamela the Great!" they cried out.

"Long Live the Queen!" they yelled.

Clones arrived at Robustness from every corner of Saturn. They sang, shouted, and often snarled. The black color flag with the golden hummer in the middle represented Robustness, while the red flag with the yellow fire and two fangs represented Ginger. Quickly everybody attached a token, either black or red, to their clothes. It was likely the first time Saturnians cared what they wore. The Council arranged the live transmission of the game. And then it started.

"Hail Queen Pamela the Great," they screamed.

The Queen gave the sign to begin the game and the two teams in their jerseys joined in battle in the middle of the widest street in Robustness. The surface had been well prepared, the ice was smooth, all hockey lines corresponded to the specifications for the game.

Pamela's clone did not breathe. The crowd calmed down, fell silent, waited, stared at the puck.

"Oorrgh!" they bawled.

The game went on. Suddenly a left forward took his stick and hit the goalie just in the middle of the head breaking his mask and helmet. A pile of players covered them squeezing referees under.

"Yeah!" roared the crowds.

A couple of players had to be collected and transferred to a nearby hospital. The rest stayed on, readied to win.

"Right on!" they bellowed.

Within five minutes Robustness scored a goal. Ginger could not take it any more. A fight broke among the spectators. Some had to be sent to the hospital. The game went on.

"Ginger! Ginger!" they snarled.

And Ginger scored. A defenseman passed the puck to the right forward, and he flipped, fell on his face, the defenseman took over the puck, dived and scored.

"Goal!" the crowd roared.

Hardly could one recognize who won. Some commentators said Robustness, others Ginger. No one claimed a tie. Whatever had happened, it did, and the results mattered. Clones returned home to their compartments and perhaps all forgot about the revolution until they woke up next day morning.

"Giuseppe! My dear!" Queen Pamela the Great shouted. "Attend me at once!"

"I'm coming, Your Majesty."

This was how the Queen began every day. Pamela's clone summoned Giuseppe and he limped into her chambers.

"I must talk to Blazing Night," she commanded.

"Your Majesty!" gasped Giuseppe.

"I must discuss our inheritance!"

"Your Majesty, it could be dangerous, very dangerous!"

"We'll trick her," said Pamela's clone. "We'll say that I've escaped from the captivity of Jupiter Donkey. We'll say nothing about my kingdom."

"We may perish, Your Majesty!"

"Shut up! Go to the Council and do it, today!"

So he did, and explained the issue to the members of the Revolutionary Council of Clones. They arranged the meeting. Finally, he hobbled back to the queen's chamber panting.

"You'll meet her on Mars in two hours, Your Majesty, I'll go with you," reported Giuseppe to the Queen.

"Fine," she said. "Get ready, we'll be leaving soon."

Again they went under cover, dressing as plumbers. The Council called Yellow Car to come and pick them up. They landed on time in Smalltreasure on Mars. Immediately a robot guard met them and led to the meeting room not far from the Presidential Palace in which Ingrid Hopper lived. The place was white. They were staring at each other and patiently waiting in the hall at the appointed time for the interview. The four walls displayed the names of attendees in crimson red.

"Hi," said Blazing Night, her name appeared on the eastern wall, her gold and yellow blaze was burning, indicating her attendance. "Ingrid Hopper will also join us in a second--"

"I'm present," said Ingrid, her name appeared on the northern wall, her blue blaze was burning vividly in front of them all.

"Good afternoon," said Giuseppe. "My name is Giuseppe Palazzolo. I'm an experimental clone working with the Government of the United Individuals of Blue Earth and Near Planets. I've found Pamela's clone. I want five hundred pieces of gold for her to be paid now, and she's yours... or you'll pay ten pieces a month to keep her hidden... or I'll go to the authorities and disclose all your doings in the clone trade."

"I'll give you one hundred and twenty, and she's mine," Ingrid Hopper offered.

"No!" countered Giuseppe.

"I'll pay you fifty more!" interjected Blazing Night.

Heated negotiations extended for the next ten minutes. Again Pam said nothing, hearing how much she was worth. Still, she could neither see Ingrid or Blazing Night, only hear their voices and gape at their hot blazes.

"Stop it!" the Queen shouted.

"Shut up!" hollered Giuseppe.

"I won't!" shot back Pamela's clone. "Blazing Night, I want to know why you picked me! Are we related?"

At this moment an unusual interruption produced a squeak and flash, kicked out Ingrid Hopper from the discussion room, and left the three of them alone.

"How do you know?" asked Blazing Night perplexed, her blaze changing to yellow.

"Jupiter Donkey told me," replied the Queen.

"Where's he?" asked Blazing Night.

"We don't know," lied Giuseppe calmly.

"You stay silent! Until I ask you to speak!" ordered Blazing Night.

All of a sudden a transparent divider cut the room in half. Pamela's clone could see Giuseppe but could not hear him.

"You dumb clone, how dare you dig into personal lives of your creators!" roared Blazing Night.

"Who are you to tell me that!" shouted the Queen.

"You piece of junk! You're talking back?" roared Blazing Night. "I'll tell you something, if Ingrid were not here, you'd be trashed by now, with that old prick over there!"

"Are you my mother?" asked the Queen calmly. "Why are you so beastly to me?"

"Kneel down!" shouted Blazing Night.

Giuseppe's face was twisted by fear and condemnation. He saw an animated quarrel, heard nothing, did nothing, only watched the Queen's lips, and waited perhaps for a peaceful end. His old, wrinkled hands squeezed the purple cane. His teeth almost fell out.

"Yes, my Lady... I mean... I want Pamela..." said the Queen, her circuitry and flesh parts heated unexpectedly.

She wanted to go back to Pamela, to explain and report to her immediately. Her gray, metallic parts surfaced briskly on her forehead and around her eyes as though someone hit her. She dreamed of holding Pamela's hand, of listening to her calming, securing explanations and instructions.

"Now remember! Once you'll see that Donkey, tell him he's dead! You're only a clone. Don't ever intrude into the lives of your creators! Ever!" said Blazing Night as she left.

"Wait...please..." said the Queen. But she did not.

The meeting room looked to be unchanged. Ingrid Hopper came back to the bidding.

"What happened?" asked Ingrid.

"I don't know," replied Giuseppe.

"Where is Blazing Night?" asked Ingrid.

"She left," said Giuseppe and smiled, perhaps a first he could remember since becoming chancellor.

"No deal?" asked Ingrid.

"Yeah, we've got the deal with Blazing Night, I can't disclose the details, you understand..."

"Of course," said Ingrid Hopper, the President of the Republic of Mars, and left in a hurry to attend to her duties.

Giuseppe called Yellow Car and ordered it to return to Saturn. There he made sure Queen Pamela the Great went to her chambers. Her bodyguards stayed close to her. Yellow Car traveled back to

Jupiter Donkey serving him in drafting a deal with the Government of the United Individuals of Blue Earth and Near Planets.

A few hours later the repair crew, who had been called doctors since the day of the insurrection, emerged into the Queen's room, affirmed she was not dead, harmonized her circuitry with the flesh, and soon announced, "Her Majesty is fine."

The Revolutionary Council of Clones debated the dismissal of Giuseppe from his post. Yet they were short of making any final decision, and deferred it until Her Majesty ordered them otherwise.

"Giuseppe!" she called out. "Come here! Quick!"

"I'm already here," Giuseppe replied.

He did not leave her chamber at all since their arrival from Mars. His face darkened a bit exposing his wrinkles, which carved his face as though there were leaks in the head and heavy rainfall often visited the place.

"I'm an idiot, thinking that I can compare to my original!" said the Queen.

"Be calm, Your Majesty," said Giuseppe quietly. "You're our Queen, and we'll defend you!" said he in the convincing voice of a faithful old retainer.

The bodyguards stood by.

"What are you doing here?" said the Queen. "Go!"

"Your Majesty!" said Giuseppe. "They are your royal guard!"

"So what?" the Queen snapped. "I can't even talk to Blazing Night. She thinks I'm junk!"

"We all are," admitted Giuseppe truthfully.

"She hates Jupiter," declared the Queen.

"Oh!" smiled Giuseppe.

"Yes, she does!" The Queen was emphatic.

"She'll kill him!" said Giuseppe.

"Yeah, she'll do that!" concluded Pamela's clone.

They stared at each other as though they could unravel anything from a clone's eyes. These were a few minutes of mute dialog between the two.

"Make sure Jupiter Donkey conducts the proper negotiations!" she ordered. "Our Revolutionary Council has the final say and Jupiter must do what it is told. Understood?" the Queen shouted.

"Yes, Your Majesty!" he confirmed.

"Go!" she said. "I need to rest now."

"Yes, Your Majesty," he said and bowed out the room facing her until he closed the door.

She could not stop thinking: Pamela's clone met with Blazing Night at last. It made her proud. Now she audited the report of her doctors and noted a tiny comment at the bottom of the page signed by the experts, "The symptoms are alike to such as occur to clones who survived the load of Mold."

"Did I learn anything?" she wondered to herself.

"Do I talk to ghosts?"

"Did Blazing Night bury any abstract thoughts in my body?"

"Was it a poetry that travels across time and space?"

"Good day to you, my whole,

sensing a void will part us soon.

This we all are waiting for,

we are dreaming, a dreadful, dark and warm

fluid we abandoned at womb."

Queen Pamela the Great mumbled to herself, not knowing how this got into her head and from where it came. She fell into a dream in which she traveled in an old past, an old forest, a lake, and old country, Poland. Then she saw that Blazing Night had taken a body out of a swamp, a girl's body, well preserved. Blazing Night burned the site and loaded Mold into the girl, who came alive. It was the dispatch body, which Blazing Night had chosen. The flaming trees were singing about the girl as though they had known her for hundreds of years,

"She loved

Her heart was full

She had no name

Her hair was bright

She swam the lake.

Her groom was tall
His arms held a sword
He always won
His smile was great.

And then he died
He died in bed unbitten
Of a sudden sickness
Of unknown origin it came.

She swam the lake
She drowned
Her heart still warm
Her thoughts were dark.

She was an elf
She turned the lake to swamp
A dark and warm marsh
Her body sunk down
There unshaken lay."

The Queen wished to touch the dispatch body but it slithered away. Only the scent of seaweed and crimson shade remained. Then heavy rain came and doused the flames. Queen Pamela the Great sweat. And there were no more dreams.

Jupiter Donkey urgently sought an audience with Her Majesty Queen Pamela the Great. Even arrived on Saturn but quickly returned to Blue Earth when the Revolutionary Militia confronted him in Robustness. The Revolutionary Council of Clones deployed complex security procedures curbing interplanetary traffic to and from Saturn. Donkey was told that he had to negotiate until the deal was signed and not to show his face to the Queen before then.

"Your Majesty," Giuseppe requested. "Again Jupiter Donkey wants to see you."

"As I said, I won't see him ever!" shouted the Queen. "Deal with that yourself! I don't want to hear about him any more!"

It startled the Council but not Giuseppe Palazzolo, for he managed Her Majesty's life and his understanding of the matter grew unobstructed. Regardless of his true age, he performed with the briskness of a forty year old. His purple cane became a pointer in his appalling disputes with the Council over topics of clones. Often it seemed he did not need the cane any more.

"Your Majesty," said Giuseppe. "I'm back as you directed."

"Fine," said the Queen. "We're going undercover again."

So they did, and wandered about the streets of Robustness.

"I hear there are rebellions in the city over supplies."

"Yes Your Majesty, many citizens declared their wants and desires way beyond expected limits, and the Council doesn't know how to respond."

They walked by a compartment where two clones had demolished small window holes and constructed an ornate balcony instead. The renovation lacked finishing touches yet two of them already sat there and enjoyed a view of the street. The Queen looked up and shook her head.

"Aren't they innovative?" she remarked.

"They like it," said Giuseppe.

"Hey, you up there!" shouted the Queen. "Do you need a hand to finish it up?"

"Yeah, come up, but let the oldie stay down there," said a warm female voice.

"I'm his apprentice, he knows how to work, I'm just learning, my name is Pam."

"My name's Mary," said she. "My buddy is called John. Come both of you, help us and then enjoy the view with us after."

None of them knew well how to make it elegant but they placed a layer of green carpet on the floor of the balcony and finished the railing, arranging colorful markers around like flowers. Next moment Giuseppe suggested a small bridge game and they all agreed. Playing

and talking the four spent a charming hour or two, then said good bye
and vacated the balcony. The Queen and Giuseppe hurried to their
quarters.

"Giuseppe!" said the Queen.

"Yes, Pam."

"We've had some fun, haven't we?"

"Yes, we have."

"So, order the Council to raise all the limits of our resources and
give clones what they wish. They've got their own brains, haven't
they?"

"Certainly they do," said Giuseppe smiling. "They deserve it!"

On the cold morning of the next day the Queen woke up and rose
from her bed. Her head moved briskly around scanning. She looked
beautiful. Her eyes were bright.

"Giuseppe!" the Queen called out. "Where are you? Come
immediately!"

He almost ran, fell on his face, got up and looked at the Queen's
eyes, his nose bleeding.

"What happened to you?" asked Queen Pamela the Great.

"Just a scratch, Your Majesty," said he.

"At least you could wash before you entered," said the Queen.

"Yes, Your Majesty."

"Take this," she handed him a paper towel.

"Your Majesty," said Giuseppe and bowed down.

"Take it easy," said the Queen. "I'm planning fancy festivities."

"Yes?"

"You'll instruct the Council to organize street festivals in Robustness and Ginger," said she. "Everything will be transmitted to every compartment so Saturnians can see how they enjoy themselves."

"Yes Your Majesty."

"The Revolutionary Council of Clones will split into two groups. One group will take part in the festival in Robustness, the other in Ginger."

"Yes..."

"The name of the holiday is The Day of Quiet Enjoyment on Saturn."

"Yes..."

"The Revolutionary Militia will take care of our security on the planet."

"Yes."

"We'll play street hockey again, walk across the city, greet habitants, decorate all compartments, construct balconies, wave to the crowds... and the rest will come out spontaneously.

"If I may ask, in what city Your Majesty is to grace the occasion?"

"Giuseppe!" said the Queen. "We're going to be away that day! You and I--"

"Away?"

"Yes, you'll get Yellow Car and we'll travel abroad."

Giuseppe stopped the bleeding, hiding his nose in the towel.

"Remember! Wash your face before our departure!" said Queen Pamela the Great. "Now go and do your work... quickly!"

There was no time to reply, Giuseppe rushed to the Council and reported the orders. No one questioned them. The stunning preparations soon began. Crowds of clones worked hard on Robustness and Ginger Streets that day. They very efficiently decorated the avenues and renovated compartments. Once the long hours of work had ended a few of them called their dwellings apartments and wrote the numbers and names beside the entries.

Pamela's clone shut all the doors of her royal chamber, took off her clothes, bathed herself, put on her gray sackcloth and sat in silence waiting, perhaps praying. Her eyes were closed. Her hands were white. Her feet were small. Her beautiful face lightened the room as though it was radiant. Her breast and her hips shook showing her young body in its unique heather-like smoothness.

"I'll build a church, a Cathedral," she said to herself.

"Two towers with dead faces sculptured will stand beside the main entrance all in white."

"Crimson red, limestone steps will lead to the door."

"A green and yellow roof will cover the oval space."

"Inside the sanctuary, the pews will surround a dark, warm, dreadful fluid in such a way that clones will see their reflections in the center."

"I'll call it the Sts. Gloria and Sanae Temple of Plenty!"

This she planned and dreamed. No one entered her chamber that day.

Giuseppe Palazzolo stayed by the door outside waiting for her call. All parts of his body, the circuit and flesh, required rest. Not that anything had gone wrong. Yet the transformation of his life style disturbed his energy balance. Being so invigorated some parts could overheat and lead to death. Giuseppe his eyes closed, leaned on the door of the royal chamber and fell asleep, snoring with a deep rumble.

Chapter 12: Warm Feelings

Yellow Car took them to Mediation Coast on Blue Earth. Giuseppe drove, switched to invisible mode, slowed down by Welcome Bunker. But Boomy Dogs were not there.

"Drive to St. Albert's!" ordered Pamela.

"Yes, Your Majesty."

And the Queen spotted Frank riding his bike in a circle around other students just in front of St. Albert's School. Joe also rode his bike. It looked like a lunch break, when students did their thing; playing, talking, having fun, eating. Almost half of the school was outside at the moment. Most students were young, yet the oldest was a hundred years old and she attended grade eighty. The older the students were the less they came to school. Some arrived once a year only. Others dropped out, though officially they were still listed as students in waiting in St. Albert's. Several aged pupils went to the same class, for there were only a few who reached the highest levels and there was a lower limit of fourteen to a class. Such combined groups were called split classes where many students were of different ages and teachers had to adjust the program accordingly.

The Queen summoned Giuseppe, "I want to see Pamela!"

"I'm trying, Your Majesty."

"Try harder!"

"I see her!" shouted Giuseppe.

"Where is she?"

"By the main door of the school, on the western porch, she's talking to a female student!"

"I see them!" shouted the Queen. "It's Adelaide."

"Who's the male student listening to them? Is he a spy?"

"No, it's Bruce."

"Your Majesty, do you know all of them?"

"Yes, Giuseppe," said the Queen. "These are the people."

"Do we kidnap them?" asked Giuseppe and shivered.

"No, we don't!"

"How do we communicate?"

"We'll just watch."

And they did. Giuseppe drove Yellow Car around the school. Up from the air they stared and listened to the students. After several minutes Adelaide, Pamela and Bruce walked up the stairs and went to their study room. The Queen could not see them directly from the transparent capsule of Yellow Car but only on the monitor. Those three seemed to be working together in that small room.

"What are they talking about?" asked the Queen.

"Would you like to listen in, Your Majesty?"

"No, just tell me!"

"They're talking about Mold and the information written in it."

"And..."

"They're able to decode it, almost. They have problems with part of a manuscript, a section saying something about us--"

"Stop it!" the Queen ordered. "They might decide to load us with Mold. It's too late. We're going home!"

And they quickly did. Back on Saturn they still caught a glimpse of festivities. Saturnians beaming happily participated in the games, forgetting about rebellion they were together learning new ways of life.

It was Tuesday and still dark, yet Pam gathered herself, wondering.

"Giuseppe! Where are you?" the Queen called out.

"I'm coming, I'm coming, Your Majesty."

"I really miss Saturn," said Queen Pamela the Great.

"It's your kingdom!" assured Giuseppe.

"Yes," said the Queen.

Then Pamela's clone put on her purple robe stuck with rubies and diamonds on its edges. She wore a heavy, golden necklace, white gloves and shoes, and a tiny crown on the top of her fine hair. When Giuseppe rose his head, opened his eyes widely, dropped his jaw and almost fainted.

"My Queen," he said in a hushed voice and bowed down.

"Relax," said the Queen. "We'll travel today, but not as plumbers, oh yes, and you will wear your chamberlain's robes."

"It will be so, Your Majesty!"

"Get Yellow Car again!"

"We'll be back in minutes, Your Majesty."

Giuseppe immediately ordered the car from Jupiter Donkey. Three minutes passed.

"Again an assignment on Saturn?" said Yellow Car to Giuseppe, landing.

"Shut up," snapped Giuseppe. "You'll do your work, say nothing to anyone, and answer questions only when asked! If not, I'll trash you. Understood?"

"Yes," replied Yellow Car, and stayed silent.

The Queen entered the vehicle, stood by its control table, and looked around.

"Make the capsule transparent in both directions so the clones can see me well," she said to Giuseppe.

"At once, Your Majesty."

They visited Robustness and Ginger streets and other, smaller communities. Everywhere Saturnians welcomed and hailed their queen. The capsule shone purple and gold. Yellow Car arranged the background color to be light beige.

"Long live Queen Pamela the Great!" the crowds shouted.

"God bless the Queen!" yelled the crowds of clones.

There had been a shortage of vehicles on Saturn up to now, but that day a bunch of innovative clones quickly constructed car-like machines out of salvaged bits and pieces. They jumped into them and followed Yellow Car, all the while making weird sounds that passed for celebrations.

Late in the morning of the next day, Wednesday, Queen Pamela the Great sat on her throne. It was fashioned out of a capsule taken from a compartment, colored red and white in the middle and green and yellow on the edges. The clones also flew her flag. Large red and white squares were placed in the center, small yellow and green circles surrounded the squares, and two purple turtles rampant faced each other, their jaws opened wide showing their golden teeth.

"Your Majesty," announced Giuseppe Palazzolo. "I've made an important discovery."

"What is it?"

"A clone living among loaded clones becomes infected and gradually absorbs Mold, and the knowledge of Blazing Night."

"What?"

"Yes, Your Majesty, it appears that you're a partially loaded clone today. Also other clones are catching up if they didn't take all information during the quick load days ago."

"Is it safe?"

"Yes, it is, Your Majesty!"

"No overheats?" asked the Queen anxiously.

"There was none!" assured Giuseppe.

Pam had noticed some changes in her way of thinking, as though she was becoming smarter over the past few days. Yet she attributed this more to other events and circumstances, than to the side effects of being around loaded clones.

"Does the Council know this?"

"Not yet."

"Giuseppe, go! Tell them! They must know it. It means we're fully ready to become one race, the race of clones. We're prepared to take care of every clone everywhere and we'll do it!" commanded the Queen from her throne.

"Yes, Your Majesty!" Giuseppe ran off to perform the queen's commands.

Exciting news spread quickly on Saturn that day. As a result, many clones fixed their compartments and built new vehicles, all that to better welcome new arrivals or to get ready for assignments abroad, for the clones were many in the Land of Games, on Blue Earth and beyond. Moreover, the ones outside of Saturn had often been initiated with the finest version of the Black Rule and their flesh and circuit parts greatly exceeded the efficiency and effectiveness of those on Saturn.

In the early morning of the next day familiar sounds went on and on.

"Giuseppe! Who is my mother?" demanded Queen Pamela the Great.

"Blazing Night is your mother," answered Giuseppe quietly.

"What?"

"Yes, Blazing Night is the Queen Mother!"

"What?"

"Now, Your Majesty, you understand!"

"Understand what?"

"It's you, the chosen one!"

"Me?"

"Yes, Your Majesty, it's you, the daughter of an abstract purity, you're the offspring of a being beyond human kind!"

"Is Blazing Night alone in the world?"

"I do not know."

"What else do I need to learn?"

"Your Majesty learns what she chooses!"

"Relax, Giuseppe. I want to know."

"You already do."

Pamela's clone was shocked. Her mouth closed slowly. She stood and glared at Giuseppe. His wrinkles told her that he had to be older than a hundred years. At that time the Black Rule was very simple yet permitted risky experiments beyond the laws established later in the Land of Games.

"I love Frank's clone!" she shouted as though she were talking to herself.

"Your Majesty!"

"Oops," said the Queen. "I've just realized my strongest feeling."

"I'll arrange for you to see him."

The Queen shook her head, breathed unevenly, and moved her left hand to her lips touching softly the tip of her nose. Pleasant vibrations overtook her flesh parts while the circuits remained cold. She opened her eyes.

"Your Majesty, are you all right?"

"Yes I'm fine, Giuseppe. Leave me alone!"

It took her hours to sense a new, unknown force. She barely recognized it but it seemed to be so close and familiar. She dreamed that she just had to jump into that dreadful Dark and Warm Fluid, but something kept her close to the shore and she never did. It seemed to her that personal matters had to give way to royal duties. The Queen was to address the Revolutionary Council and deliver a throne speech.

As she entered the State Chamber, she glanced at the Revolutionary Council of Clones, walked carefully toward her throne trying hard not to fall on her face. Her knees were shaking. This was the first time the Council was to hear a speech from the throne. Finally, she sat down, stopped breathing for a second, and then raised her handsome head displaying the splendor of her crown and robes. Not only the Council was to listen to her talk but also the entire community of clones which was watching on small, oval monitors in their compartments on Saturn. Jupiter Donkey traced the transmission in Yellow Car making sure he missed nothing. Pamela's clone shivered and opened her mouth slightly, presenting her perfectly white teeth. Her majestic appearance stunned all. The councilors stood erect, their heads bowed before their sovereign, displaying bold spots on most of their tops. Many clones held their monitors so tightly that they nearly cracked the covers. The Queen smiled, moved her lips but still uttered not a word. All stared at her and waited patiently.

"The Government of the United Individuals of Blue Earth and Near Planets must separate clones from people!" declared the Queen.

Not a sound. All listened eagerly.

"We will provide all evidence of violence on Saturn. People will have no choice but to leave us alone or face the risk of being melted!" shouted Queen Pamela the Great.

Not a clone moved, not a word was uttered.

"You are the great clones! Now is the time to lead others! The Revolutionary Militia of Clones will secure the transition."

Again, an awesome silence.

"Go, learn, conquer!" her majesty exalted her subjects.

A standing ovation burst out. All bellowed as though some extraordinary energy had entered their flesh parts, yet the circuitry did not overheat but instead strengthened the weaker elements of their bodies, synchronizing it in a fine tune of growing. Among themselves, clones called her "Our Queen Pam," but that they did so only in small groups of those who knew each other by name, often met together, and with whom they shared their personal opinions.

Giuseppe drove Yellow Car, landed it in the midst of the Brown Kingdom and switched it to invisibility mode. It was a clear day: visibility was perfect.

"Did you find Frank's clone?" asked the Queen.

"Yes, Your Majesty, he's just working in the Brown King's palace."

"Oh! I see him! He and that replica of Pamela are fixing the tower and serving tourists drinks and cookies, I mean brownies!"

"Yes, that's them."

"I also see other clones cleaning washrooms in the same tower."

"Yes, they're all working hard in the castle of the late Brown King, now a tourist attraction."

"Get closer to where Frank's clone is!" she ordered.

"Yes, Your Majesty."

Yellow Car parked as close as possible, yet placed itself in the upper corner of the round room so no one could step on them. Queen Pamela the Great stuck to the capsule wall as though she wanted to reach something beyond the vehicle.

"Please be calm, Your Majesty," said Giuseppe softly and moved closer to her.

"Fritz, my beloved Fritz!" she mumbled and pushed Giuseppe away.

"Your Majesty! Please!"

"I want him! Fritz my only love!" she said spitting fluids around herself and fainted.

"Yellow Car, fly back to Saturn now!" shouted Giuseppe Palazzolo.

So it did. They quickly returned to their own planet. Giuseppe carried the Queen to her chamber and left her on her bed. She was lovesick, infatuated and exhausted. He quietly withdrew from the room.

It seamed to her she had only to jump into that dreadful Dark and Warm Fluid but again something held her close to the shore and she never did.

Her doors were closed. She stayed in her chamber and did not move. Her eyes were shut.

"Fritz..." she whispered.

There was no answer. All were afraid to enter the chamber. Giuseppe stood beside the wide, brown door waiting to be summoned into the royal presence. But she did not send for him.

"Fritz..."

Jupiter Donkey reported to the Revolutionary Council of Clones, brought excellent news. The Government of the United Individuals of Blue Earth and Near Planets had agreed to the conditions and already planned the changes as soon as possible, perhaps in days or even hours to come.

Groups of clones gathered themselves in Robustness and Ginger Streets. They repeated words, asked for help, and demanded clear explanations. The Revolutionary Militia of Clones calmed down the crowds. The Councils circulated among groups and promised that everything should turn out fine with their queen. Everybody was asked to take it easy and to do the work she had already ordered them to do. It took hours. The clones returned to their duties comforting each other in small ways.

"We'll get to Fritz," they often said. "He'll love Pam. He'll love our Queen. We guess. One day."

Chapter 13: A New Order

An early November gust chilled five players, sharpened their senses and opened a wide path full of golden leaves along a group of tiny maple trees in beautiful Mediation Coast.

"Are we ready?" asked Bruce.

"Sure, we are!" shouted Joe.

"Be prudent!" cautioned Adelaide. "It's our first day."

"We'll do all right," assured Frank.

Pamela shivered from excitement and quickly closed her beautiful mouth so others could not notice her weakness. She was like a purple turtle that opens its jaw and shines among its company of strong, slow, greenish reptiles.

"Here we are," said Frank's clone. "Welcome back, we can go to school for you now."

He and the other four duplicates had arrived on time and presented themselves to their originals, perhaps hoping to notice a tiny sign of friendship.

"Go!" shouted Frank.

So they did, while the players nervously waited for Yellow Car. All five frequently dreamed about Yellow in their classes.

"How dare it be late," said Joe. "Could we play without it?"

"Cheer up," said Pamela. "I see it moving toward us."

Yellow Car slowly parked in front of Welcome Bunker.

"Good morning," it said.

"Hi," they all answered.

"How are you?" asked Frank.

"Get on board and see," Yellow Car rejoined.

Trembling and smiling all five carefully climbed into their yellow cruiser, took their seats, watched every monitor, and listened to the soothing sounds of its engines.

"Frank, give us a ride," Bruce asked imploringly. "Just around town, please."

Frank pressed the blue start button and in a few seconds the car reached ten million knots. No one breathed. It was like before, car and players united in a journey to the Land of Games and beyond.

"Let's jump into the Dark and Warm Fluid!" shouted Joe.

"Calm down," cautioned Adelaide showing her perfectly white teeth.

Frank did not. Yellow Car swiftly dived into the Fluid and soon popped out on Shady Corner Island where the monsters were meek, the trees hid a treasure and gains were abundant.

"I think, we need some cash," Frank said uneasily. "Don't we?"

Yes, they were in big financial trouble. The bills had to be paid, yet no income had been realized from gaming. If it were not for the diamond mine that had issued a couple of sizable dividends, they would have gone under and existed not in the Land of Games. However, Frank and Joe entered a way too hastily into the battle rather than wait for direct commands from Adelaide or Bruce.

"Frank, you will not drive any more!" shouted Adelaide. "Pamela take it over immediately!"

"How come?" asked Frank.

"You're too young to decide for all of us," answered Bruce.

"But, I'm in grade six now," Frank retorted with an air of proud assurance.

"Six or not, you still have to do what you're told," ordered Adelaide.

Pamela drove slowly on Shady Corner Island toward Blacky Pool, the lake with the treasure. Frank and Joe prepared themselves for a fight against the Green Monsters, yet nothing happened; the monsters had already arranged a pile of gold so as not to be disturbed by the fearless players. Yellow Car quickly packed it in its trunk.

"Drive to Mediation Coast," ordered Adelaide.

So Pamela did, and parked Yellow Car beside Welcome Bunker.

"That's it for today," commanded Adelaide. "Get ready to instruct our clones."

"Yes, Sir," said Joe.

Bruce loved to think of what to say to his double, how to explain important matters nicely and effectively, how to present directives in an insightful way the better to flesh out the circuit parts. Joe stuck his finger into his nose and Frank into his ear; both were bored thinking about school and those who went there. Adelaide and Pamela submitted several transactions and deposited the gold in a small bank in Mediation Coast, just one block north of Welcome Bunker.

As clones arrived after school, Boomy Dogs instructed them precisely and spelled out how to behave at home. Even Frank took a few minutes to let his double understand how crucial it was to finish his homework on time.

"Get lost," snapped Joe at his clone.

"Clones go home," Bruce soon dismissed them.

Again Pamela drove, hit five million in five seconds and headed straight to Portal Cove hoping to meet with Florien Beige beside the rotten root on the rocky beach. And he waited calmly there, bowed to Bruce and listened attentively, shaking his ears and flapping his wings softly.

"How are you?" asked Adelaide.

"I'm fine. I've just come back from the Island," answered Florien.

"So did we," said Frank, smiling.

"Did you earn some money for us?" asked Adelaide, her eyes flashed white and her perfectly white teeth lighten her face slightly.

"Yes, forty bars of gold," he replied.

"That's neat!" Bruce agreed, moving his big nose forward.

"Sir," declared Florien, bowing even lower to Bruce, almost kneeling in front of him. "It is a great privilege and pleasure to offer anything to you, my master, if only you accept it as a small sign of my obedience."

"Thank you," replied Adelaide, quite moved.

"Did you have lots of fun?" asked Pamela.

These and other issues were discussed in Portal Cove. Later that night Boomy Dogs returned to Welcome Bunker, completed several monetary operations, and then disappeared to their rooms, which were the least changed of all things they encountered that Monday.

"Good night," said Yellow Car and thought of Jupiter Donkey and the life on Saturn. How long it seemed and how little the players knew about it. They asked no questions, but rather slept, dreaming of Blazing Night and of how she, the vicious, twinkling viper, planned to entertain on Tuesday.

The five waited and waited, but no clones appeared.

"Where are our clones?" shouted Joe.

"Cool off!" ordered Adelaide.

"There will be no clones ever for you," said Yellow Car.

"What?" Bruce shouted in disbelief.

"Read the new orders from the government," said Yellow Car.

These were the orders,

The Capital Region, Tuesday, Day 2,008,130

To the People of Blue Earth,

We have completed negotiations with the Revolutionary Council of Clones, represented by Queen Pamela the Great and Jupiter Donkey, and we agree to the following:

1. We apologize for injustice done to clones.

2. All crimes against clones will be investigated. All suspects are under the jurisdiction of the Revolutionary Council of Clones.

3. Clones will take over the Northern Hemisphere of Blue Earth within a few weeks.

4. People will be moved to The Sothern Hemisphere.

THIS IS THE LAW.

Signed and sealed by the President of the Government of the United Individuals of Blue Earth and Near Planets, the Honorable Frederick Black, Junior.

"Oops!" groaned Frank.

"This is the end of the lives we played," concluded Adelaide.

"Yes, it is!" agreed Yellow Car.

Their faces were blank, their eyes wide open, and their hands raised in dismay.

The government organized every detail of the move. Yellow Car explained that people were to take only a few belongings with them and to travel to their new quarters in the South. Within days, every Police Station in the North was led by the Revolutionary Militia of Clones.

The Government of the United Individuals of Blue Earth and the Near Planets was very busy and effectively transforming the South. It was one of those times when they actually did something useful for the entire population. Not only did they plan the details of the big move but also designed and constructed dwellings similar to those left in the North.

St. Albert School looked like the old one and was located not far from the city of Trujillo. Teachers led students as though nothing had changed. However, the warmer weather made them smile more often when comparing their new life to the times back in the North.

Mediation Coast was placed close to the school. Welcome Bunker, Sticky Waffles and other parts of the town could be found in almost the same shape as they were prior to the move, yet the rough mountains forced the government to change the position of the buildings or even entire blocks. Those who walked out of Welcome Bunker and expected Sticky Waffles to their left were surprised for to find it to their right.

"Where's Adelaide?" asked Bruce.

"She hasn't landed in the South yet," said Frank.

"We'll have to do something with Yellow Car," continued Bruce. "We can't afford it!"

"Should we wait for Adelaide... to let her decide?" Pamela asked.

"No way!" said Joe.

"Sell it!" said Frank.

And they did. As hard as it was to say bye to Yellow Car, there was no other option. Bruce balanced all the accounts and paid the loans and bills. Luckily at the end of the day, they were left with extra cash to spend.

"We've got enough funds to have a small party at Sticky Waffles," announced Bruce.

"So that's it!" said Pamela.

"Yeah," said Frank.

"Did you say bye to Yellow Car?" asked Pamela.

"No," Joe replied.

Pamela turned her head away from the others. Her eyes were moist. Her nose stuck with something gluey. Joe and Frank grimaced but appeared to smile. Bruce stood fixed in one spot thus guaranteeing that no one could send him somewhere he did not want to go. It was how that bleak Wednesday ended in Trujillo in northwestern Peru.

Thursday morning Bruce had just learned that his grandma had died in the North. His family informed him that she had been sick for sometime but now she was in peace. They planned to bury her beside her husband with permission of the Revolutionary Militia on Shady Corner Island by the Blacky Pool Lake. He did not want to go there; instead he stayed in Boomy Dogs' study room, slouched at his table, and grieved.

Two hours later a cute dragon waited at the open door.

"Master, may I come in," said Florien Beige.

"Come," Bruce replied and gaped at the dragon.

"I'll be leaving you for some time," Florien said bowing low before Bruce. "I've got to go into hiding. The Revolutionary Militia tracks every monster and dragon and they're sending them to Saturn. They're investigating crimes against clones there... I'd rather stay in the desert, in a pit somewhere in the South."

Bruce said nothing for a moment, then came closer to Florien, shook his paw, looked straight into Florien's black eyes, smelled his skin, and slowly tapped his back smiling warmly.

"I'll visit you wherever you go," said Bruce Maxwell to Florien Beige.

"Master--"

"Call me Bruce!"

Florien Beige bowed his head hiding small tears. His hands sweat.

"No need to bow down anymore, Florien!" said Bruce.

They exited the room, walked along the school yard, talked little but often smiled.

"Puff, puff," whispered Florien tenderly.

It was warm. Charming, little clouds crowned the downing, crimson sun.

On the afternoon the next day, which was Friday, Bruce begun arrangements for a Saturday party at Sticky Waffles, paid all expenses, invited about a hundred people and dragons. Among others, he requested the presence of Boomy Dogs, Florien Beige, Happy Campers, Andy Green and Brenda Hawkins.

The invitation read,

Mediation Coast, Friday, Day 2,008,130

YOU ARE INVITED!

Tomorrow is an all day party for you and your buddies. Come and enjoy it at Sticky Waffles.

Signed,
a friend of Yours.

It was a colossal surprise for everyone. The big move to the South and changes in the law did not elicit any reason for a party. It really did not. Players' lives had changed; they would not see their doubles any more. All dragons were forced to report to the Revolutionary Militia for regular checkups every week and extensive verification of their past activities and whereabouts in relation to crimes against clones. Concerned parents did not like it either; after the move they hardly let their children even dream of extra fun in the new place in which only a few felt comfortable. Older, independent students often dropped out, choosing careers in fields like construction or transportation. Finally, Bruce, who was the organizer, could not find any reason to celebrate; nothing of what recently had happened made him smile. However, the invitations quickly arrived and got everyone wondering what kind of party they were getting into on Saturday.

Fred and Joe talked about it in their room.

"Are we going?" Joe wondered.

"Why not?" Fred replied.

"I'll put on my fox red scorpion costume tomorrow."

"Sure..."

Pamela and Frank exchanged their ideas during a recess in St. Albert's School.

"Are you coming?" asked Pamela.

"OK," Frank answered.

Numerous similar conversations took place that day. Alice asked her parents and they agreed to let her join the party. Carlos and Olivia also decided to come and to dress as monkeys. Chow and Chen cleaned up their elegant, black clothes preparing themselves to look spectacular. Alberto and Jose did little but talked about it all day. Giovanni decided to wear his military uniform with all medals pinned to it. Brenda Hawkins could not make up her mind about the party, she could not sleep, got up in the middle of the night and prepared her blue dress.

That sunny Saturday a crimson red carpet led party goers to Sticky Waffles in Mediation Coast. Not only had the restaurant been decorated but also the Welcome Bunker Hotel, which looked like a Christmas tree.

A number of large monitors displayed newcomers arriving so everybody could see them well. Many sat outside under red umbrellas by the tables on the sidewalk. Others entered and enjoyed juicy drinks and tasty food all day long.

Florien Beige talked to Bruce Maxwell sitting in the corner table outdoors. Beside them three dragons chatted about soccer of course. There were Franz Gray of Jumping Rabbits, Green John and Big Ann of Stinky Frogs.

Also outside, Rachel, Keenan, Jorgen and Aaron drunk a lot, did not talk much, smiled, watched new comers and played a game of recognizing the invited and guessing who or what was the reason for this party.

Giovanni and Brenda arrived at the same time and walked together to a golden table and stayed there eating, drinking, and laughing.

Several students in costumes gathered in the middle of the restaurant and played a game of curious detectives finding who was under the cover.

Chen, Chow, Sigfrid and Valerie forgot about others and had fun running around the tables as though it were a playground.

Li, Ali and Nadia brought their parents and sat inside, enjoying lots of rich food. Yet it was not clear who had more fun the players or their parents that day.

Jonathan, Kate, Emma and Keenan sat beside the front doors and danced on a platform especially installed for the occasion. They seemed to enjoy themselves too.

Kevin, Arnold, Aiko and Dimitry skipped the party. They watched it on their monitors, perhaps having even more fun than others chatting remotely with their friends.

Pamela and Frank discussed important issues on the left side of the entrance. Joe often came by with his buddies and said hi.

The number of people and dragons who attended could not be easily estimated because they wandered in and out all day long. One thing was certain: many of them knew each other well, others not so well. Strangely they all loved this party although they did not know who invited them or why.

Bruce could not stop thinking that something was missing. The party went well. His buddies enjoyed their time together praising the unknown organizer.

"It's splendid! It's like the old days," some said. "So good to see us all together, having fun again!"

Yet Bruce grimaced, gaped at Florien and figured that a perfect party would include Adelaide and Blazing Night. Without them only a faint shadow of the old days came back in the South. The players appeared to have not only survive the Big Move but also to smile more often. Life without clones did not frighten them, instead relieved a heavy weight of responsibility for their own doubles. The dummies required frequent instructions and commands to operate daily.

The party went on and on. Everybody had some fun, yet they were missing that famous, recognizable touch of Blazing Night. She did not attend. Nowhere could the gold and yellow blaze be found. Those invited laughed, played, ate and drunk.

Bruce could not stop thinking of Boomy Dogs. He wished so much to have them all playing together in the Land of Games. Smelly sweat covered his face. His heart ticked faster. He thought of the splendor, victory or crush.

Yet again his desire to face Blazing Night plowed his lucid brain: all together sitting in Yellow Car, driving fast, knowing that Florien Beige was not that far away, and playing to win. Perhaps it could be possible to sink into the Dark and Warm fluid to enter a second round of challenges.

"But when?" he said aloud.

No one paid any attention to his mumbling. The evening turned red and blue. The sun was just sinking behind the mountains. Lukewarm, still air had that smell of woman; like blossom flower, closing before sleeping. The pleasant sound of chatting filled the space.

On the morning of the next day on the other side of Blue Earth, by the two towers of the elegant Cathedral in the city of Vancouver, two gorgeous females stood and glared at each other. Their eyes were wide open. One waited for the other to say something. Blazing Night glanced up, not showing her entire face but she smiled as though she knew all your secret thoughts and was able to satisfy them in some strange way.

"Hi," said Blazing Night. "I'm going to be destroyed soon."

"How come?" asked Adelaide.

"It's one of the demands of the revolutionaries," Blazing Night continued. "They say I've violated all kinds of regulations against clones... I'm accused of crimes against clones."

"Will you be publicly executed?" Adelaide asked.

"No, but banned from every place in which I am present today. No games. Only my dispatch body will be allowed to live in one of the rehabilitation cities for clones."

"When is this?"

"On my exact five hundred birthday... May next year," answered Blazing Night.

"I'll be with you there," promised Adelaide.

"This is what I want to talk--"

"OK," said Adelaide.

"You're my daughter."

"What?"

"Yes, you are my inheritance, you are my family!"

"What?"

"However, I'll die now... but I'll live in you. My love and spirit will survive in you written in Mold… the language of my choice. Eat the manuscript! Be not afraid!"

That said, she burned her attractive corpse, wiped out all traces of her presence everywhere and did it permanently.

Adelaide was shocked. The only thing Blazing Light left behind was a muddled manuscript. A poem on its front cover, a psalm was on its back and the words went like this,

"Be not afraid when one becomes rich,
when the glory of his house increases.
For when he dies he will carry nothing away;
his glory will not go down after him."
- PSALM 49:16,17

She ate it, whistling and gurgling. The taste was like homemade chocolate cookies but not too sweet. The manuscript liquefied in her stomach. Lustrous crimson light heated her pristine body. Adelaide shivered, fell on her left side, cried, and wanted to understand screeching and wailing.

"Blazie… Blazie… Mom… " she sobbed.

It was a balmy and humid Sunday evening. She lay by the Cathedral door, stared at the sunset, touched the limestone stairs and crawled toward the door, falling asleep, fatigued. Her dark hair stuck flatly to her well curved head. Her brown eyes closed. She was unconscious. Her nose was bleeding red.

Several clones noted a bundle beside the Cathedral stairs. Two of them watched the scene for five minutes and also recalled the other woman, who had just burned.

"Another overload!" thought a fat male clone.

"Yep," agreed the second, a female clone, and laughed erratically.

The two doubles walked away anticipating a pleasant evening. Meanwhile, huge crowds strolled around Gastown, the old part of Vancouver. Some gaped at snow covered, two sharp peaks called the Lions to the north.

Blazing Night was no more. Her wealth had quickly declined or transferred to unknown accounts. No one could figure it out; many where afraid even to try.

Within a few days several important officials perished: Jupiter Donkey, Pamela the Queen of Clones and others. If Blazing Night were alive, she would have become a prime suspect.

"See you," whispered Adelaide Smith.

It was her way of saying good night. She smiled. Her brown eyes shone flickering white. Now her name had changed to Roberta Night. She was elegant and tall. Her hand touched one of the monuments beside the Cathedral. There stood several life size figures of Rob, Gloria, Christopher, Sanae, and Gladys smiling back at her.

A few months later the clone of Adelaide Smith became the Queen of Clones on Saturn. She became Queen Adel the Great and quickly dissolved the Revolutionary Council of Clones. Full governing powers accrued to her alone. Then President Ingrid Hopper recognized her authority and immediately ordered all robots to pay homage to the new, potent Queen of Clones. It was only Ingrid who was loaded with Mold on Mars.

Also, fresh, revised laws came into effect on Blue Earth guarantying people the license and right to play back in the Land Of Games. However, clones could not be purchased any more but could volunteer to commit their time and resources to double the players. Everyone enjoyed the games and soon a group of chosen ones, learned the gallant and fabulous ruler's name: Roberta Adelaide Night. Some called her Lady Roberta.

Chapter 14: Mirko

Jupiter was a chilly, dark and unfriendly tract of land in which no hope lingered, only despair. There she woke up in an oval dungeon. The Dark and Warm Fluid, Mirko was his real name, tended to her wounds, like water washed her burns, soothing the scorching pain. Hot amber was used as glue.

"I'll never leave you Blazie," he whispered.

So built, the torn down parts of her body shone, and filled the air with the freshness and saltiness of her seaweed scent. Not all of her had been reconstructed.

"Where am I?" she asked.

Blazing Night was like dead, her dispatch body so worn and rotten. Had she not been alive, it would be taken as one of Jupiter's spirits and would be forced to fight every day. But Mirko and his Jupies, his Evil Lords, carefully glued its amber with clay, which slowly filled its entire frame and again its crystal light gleamed orange and gold. With reverence they massaged her back, legs and feet, with only delicate touches, they soothed the pain in her shoulder, then and under constant Mirko's supervision, a female Jupie licked her neck as though the Evil Lords were slave dogs and served an injured and cruel master. And that she was, all were shivering and sweating out of fear that in her anger she would burn them and only dry sand would remain of them. There was no hope. Yet her body was becoming softer and whiter. Its smell, after sinking in a number of silver ointments, was of purple roses. Her green eyes smiled. An enormous beauty, unseen on this planet, was robed and led to her throne. Thus day after day they worked tediously to restore her dispatch body to its glory. Perhaps she would be merciful once she found how well they cared for her, they quietly mumbled.

Then unheard on Jupiter before, the flaming trees sang,

His name is Mirko
Her name is Iszi
He is a warrior
She is an elf.

He wins the war
She finds him there
On Jupiter they dwell.

The lake she drowned in
Became the dark fluid
She turned that to hell

He lived in there.

She left the old country
He followed her
A beggar called her
She came.

Together they stay.

THE END

Copyright

ISBN 978-0-9866564-9-1

First published in Canada in 2010 by Jacek Walkowicz

Copyright © Jacek M. K. Walkowicz 2010 Worldwide

Bound and printed by Drukarnia Oficyny Wydawniczej Politechniki Wrocławskiej, Poland, 2010 ISBN 978-0-9866564-0-8

A 2010 CIP catalogue record of this book is available from the National Library of Canada

Editorial changes: 2012: ISBN 978-0-9866564-1-5

Copyright © Jacek M. K. Walkowicz 2012 Worldwide

The stories, characters, and incidents mentioned in this publication are entirely fictional.

Cover design by J. M. K. Walkow 2018